Sustainable Space Tourism

I0127920

THE FUTURE OF TOURISM

Series Editors: **Ian Yeoman**, *Victoria University of Wellington, New Zealand* and **Una McMahon-Beattie**, *Ulster University, Northern Ireland, UK*

Some would say that the only certainties are birth and death; everything else that happens in between is uncertain. Uncertainty stems from risk, a lack of understanding or a lack of familiarity. Whether it is political instability, autonomous transport, hypersonic travel or peak oil, the future of tourism is full of uncertainty but it can be explained or imagined through trend analysis, economic forecasting or scenario planning.

This new book series, The Future of Tourism, sets out to address the challenges and unexplained futures of tourism, events and hospitality. By addressing the big questions of change, examining new theories and frameworks or critical issues pertaining to research or industry, the series will stretch your understanding and generate dialogue about the future. By adopting a multidisciplinary perspective, be it through science fiction or computer-generated equilibrium modelling of tourism economies, the series will explain and structure the future – to help researchers, managers and students understand how futures could occur. The series welcomes proposals on emerging trends and critical issues across the tourism industry and research. All proposals must emphasise the future and be embedded in research.

All books in this series are externally peer-reviewed.

Full details of all the books in this series and of all our other publications can be found on http://www.channelviewpublications.com, or by writing to Channel View Publications, St Nicholas House, 31-34 High Street, Bristol BS1 2AW, UK.

THE FUTURE OF TOURISM: 3

Sustainable Space Tourism

An Introduction

Annette Toivonen

CHANNEL VIEW PUBLICATIONS
Bristol • Blue Ridge Summit

To my son Maximilian – may the force be with your generation

DOI https://doi.org/10.21832/TOIVON8021
Library of Congress Cataloging in Publication Data
A catalog record for this book is available from the Library of Congress.
Names: Toivonen, Annette - author.
Title: Sustainable Space Tourism: An Introduction / Annette Toivonen.
Description: Bristol, UK; Blue Ridge Summit, PA : Channel View Publications,
    2021. | Series: The Future of Tourism: 3 | Includes bibliographical references and
    index. | Summary: "This book explores the relationship between space tourism
    and the discourse in sustainability and futures research. It offers comprehensive
    information on the current understanding of the space tourism industry and
    assesses the possible impacts of space tourism on the environment, economics,
    legislation and society"—Provided by publisher.
Identifiers: LCCN 2020028368 (print) | LCCN 2020028369 (ebook) |
    ISBN 9781845418014 (paperback) | ISBN 9781845418021 (hardback) |
    ISBN 9781845418038 (pdf) | ISBN 9781845418045 (epub) | ISBN 9781845418052
    (kindle edition) Subjects: LCSH: Space tourism. | Tourism—Environmental aspects.
Classification: LCC TL794.7 .T65 2021 (print) | LCC TL794.7 (ebook) |
    DDC 338.4/7910919—dc23 LC record available at https://lccn.loc.gov/2020028368
LC ebook record available at https://lccn.loc.gov/2020028369

British Library Cataloguing in Publication Data
A catalogue entry for this book is available from the British Library.

ISBN-13: 978-1-84541-802-1 (hbk)
ISBN-13: 978-1-84541-801-4 (pbk)

Channel View Publications
UK: St Nicholas House, 31-34 High Street, Bristol, BS1 2AW, UK.
USA: NBN, Blue Ridge Summit, PA, USA.

Website: www.channelviewpublications.com
Twitter: Channel_View
Facebook: https://www.facebook.com/channelviewpublications
Blog: www.channelviewpublications.wordpress.com

The policy of Multilingual Matters/Channel View Publications is to use papers
that are natural, renewable and recyclable products, made from wood grown in
sustainable forests. In the manufacturing process of our books, and to further
support our policy, preference is given to printers that have FSC and PEFC Chain
of Custody certification. The FSC and/or PEFC logos will appear on those books
where full certification has been granted to the printer concerned.

Typeset by Deanta Global Publishing Services, Chennai, India

Contents

Figures and Boxes

Figures

Boxes

Preface

For generations, space exploration simply represented a wishful dream, until it finally became a reality in the 1950s space race between the United States and the former Soviet Union. The missions of NASA, such as the Apollo project, began to generate interest in space research and the idea of space tourism. Inspired by this change in society, Hollywood producers started to create space-related science fiction movies that enhanced audiences' visualisations and their desire to visit space. It was commonly thought that, within just a few decades, such a form of travel would be made available to the public. However, no suborbital space tourism occurred until 2018 when a Virgin Galactic crewed test vehicle, built for commercial passenger service, managed to safely reach the altitude where space begins, according to NASA's specification (Virgin Galactic, 2020). Space hype was also accelerated by an almost science fiction-style live broadcast of a Tesla car with a human-like passenger on a SpaceX Falcon Heavy rocket being launched into space (SpaceX, 2020).

Duval and Hall (2015: 450) define space tourism as 'the temporary movement of people for non-military reasons beyond the Earth's atmosphere'. There are various different types of space tourism. They can be broken down into terrestrial space tourism such as Earth-based activities and cyberspace tourism; atmospheric and low-Earth orbit tourism; astrotourism, referring to experiences beyond Earth's orbit; and lunar and Mars experiences (Carter *et al.*, 2015; Cater, 2019). Currently, the Karman line defines the boundary between Earth's atmosphere and outer space, with the Federation Aeronautique Internationale specifying it as an altitude of 100 kilometres, while NASA defines it as 80 kilometres. This complicates regulatory measures as aircraft and spacecraft fall under different global treaties and demonstrates the negative impact of the currently almost non-existent global space legislation (FAI, 2018; NASA, 2019a).

Only a handful of people have thus far visited space as paying tourists in orbital space flight, compared to about 550 professionals, mostly US astronauts. Dennis Tito became the first paying space tourist in 2001, travelling on a Russian Soyuz rocket to the International Space Station.

He preferred, however, to be classified as an 'independent researcher' rather than a 'space tourist', as his week-long $20 million stay had involved six months of astronaut training and hours of physical exercise (Wall, 2011).

Rapid developments in technology during the 21st century have accelerated predictions for the beginning of the future space tourism industry, and space tourism is likely to generate significant public enthusiasm for space exploration as soon as it becomes safe and affordable (Anderson, 2005). A new space race is on between private space tourism companies, such as Virgin Galactic, Blue Origin and SpaceX, all owned by powerful leaders, hoping to gain the status of being the world's first fully operating space tourism company. Initially, providers will offer space jumps to experience weightlessness and witness the curvature of Earth from space. SpaceX's longer-term target is to eventually colonise Mars (SpaceX, 2019a).

In 2019, NASA also joined the 'race' by announcing plans to open the International Space Station to tourists in 2020 as a means to gain more funding for other space exploration projects in the future (NASA, 2020b). The emergence of the new space economy, which provides a mixture of governmental and commercial space programmes, has the potential to create strong competition due to the diversity of contractors, offering the tools for exponential economic growth. It appears that the main goal of space tourism is to develop it to become accessible to the masses; however, the initial high cost will limit the pioneering stage to only healthy, wealthy travellers. When the space tourism industry matures, space trips could become similar to travel industry holiday packages for a range of lifestyles and budgets – the only difference being that they take place beyond Earth.

Space tourism will become the latest addition to the tourism industry in an era when the human anthropogenic influence and modern mobility have been the primary causes of the destruction of the biosphere and currently dominate sustainability discourse (Spector *et al.*, 2017). The principle of sustainable development should become relevant in planning the future space industry as the global climate change crisis makes a corporate environmental approach a necessity. Travelling beyond the biosphere has a significant impact on the Earth's climate; the launching stage of space travel pollutes by creating emissions, dust and noise in the local area, and it has also been estimated that 1000 space launches produce the equivalent carbon footprint of an entire year of global aviation (Ross *et al.*, 2010). Another prominent environmental problem connected with new space activities is the creation of space debris (Viikari, 2007).

The question of sustainability also arises within space development, with potential implications for those who are excluded or otherwise left behind (Spector & Higham, 2019a). Offering another perspective,

Carter *et al.* (2015: 457) claim that 'space tourism presents an important philosophical challenge that can be harnessed for sustainability, forcing participants to consider their place in the universe, relationship to other beings, and especially concepts of time'. Cosmologist Stephen Hawking (2010) viewed extending out into space as our only chance of long-term survival, though Peeters (2018) has critiqued the interpretation of sustainable development as something to simply save *Homo sapiens*. There are also several ethical and legal questions around sustainability, such as at what costs (to human society) should the private space sector be allowed to pursue its goals, who has the right to determine ownership in space and what should the ethical responsibilities of private space entrepreneurs be both on Earth and in outer space?

Despite its relatively short existence, the commercial space tourism industry has already presented operational-level sustainability in a way that has not been seen in the traditional governmental-led space industry. For example, in 2018, SpaceX achieved the world's first repeat flight of an orbital class rocket, presenting a historic milestone for full rocket reusability. Blue Origin plans to share the physical launch site at the once exclusively government-owned domain at Cape Canaveral. Also, at the governmental level, NASA has adopted a more sustainable approach to space flights in the Earth's orbit by developing mitigation standards aimed at reducing orbital debris (Sustainability 101, 2020). Overall, one could speculate that space tourism in the future could contribute to Earth's sustainable development by creating a new industry generating wealth and jobs, providing the technology required for low-cost access to space and hence enabling the use of solar system resources such as solar power, ultimately allowing humans to create space colonies as an insurance policy. However, it is suggested that more attention is required in order to 'develop a coherent, long-term conceptualisation of modern mobility for sustainability' (Spector *et al.*, 2017).

The Purpose of this Book

Despite the fact that commercial space tourism began to emerge after the millennium and the full operational start of this economically lucrative industry is just around the corner, the academic discourse and other written literature on space tourism, and especially sustainable space tourism, are still scarce. In the future, more academic and scholarly input will be needed to understand the far-reaching impact of space tourism on the structures of society. This book aims to address this important gap, with the following objectives:

(1) To provide comprehensive information on the current understanding of the space tourism industry.

(2) To explore the relations between space tourism and the discourse on sustainability and futures research.
(3) To introduce tools for sustainable futures forecasting.
(4) To assess the possible impacts of space tourism on the environment, economics and legislation, as well as the social surroundings.
(5) To provide visions for the future of sustainable space tourism gathered from empirical research.

The rationale is to provide readers from various backgrounds with a deeper understanding of space tourism and future sustainability discourse. This book aims to engage the general public interested in the idea of space; students of tourism, sustainability or futures studies; individual space tourist 'hopefuls'; space tourism industry operators; and various policy regulators, in the hope that it will encourage more global dialogue and critical examination of aspects related to future sustainability – whether on the Earth or in space.

The literature on space tourism has, in the past, mainly concentrated on exploring the phenomena of space from the angles of sci-fi fantasies, technical science developments or specialists', such as astronauts', experiences. Investigating new developments of the future, which do not yet even operate, is a challenging task. For this reason, the content of this book is partly based on the empirical findings of the author, collected during her world pioneering PhD research on sustainable space tourism.

Some of the existing, however rare, academic literature on space tourism, sustainable development and futures forecasting has been utilised to provide readers with a base context for understanding the theoretical concepts of this book. Space tourism research still exists more in the format of scientific articles rather than books, and some of the previous findings have greatly helped to deepen the context of the chapters. Of particular note are: Abitzsch's (1996) visions of future space tourism prospects; Ashford's (2002, 2009) approaches to space transportation; Cohen's (2017) paradoxes on space tourism; Cole's (2015) prospects for space tourism planning; Collins' (1994, 2003) and Collins and Autino's (2010) market demand and other contributions to space tourism; Fawkes' (2007) definitions on sustainability in space tourism; Duval and Hall's (2015) pioneering view of sustainable space tourism; Marsh's (2006) ethical and medical dilemmas of space tourism; Ross' (2002), Ross and Sheaffer's (2010) and Ross et al.'s (2014) rocket emissions analysis; Spector and Higham (2019a) and Spector et al.'s (2017) Anthropocene relations to space tourism; and Toivonen's (2017) framework for sustainable planning for the future in space.

Online, various official websites were referred to for up-to-date information on statistics and the latest achievements in both the governmental and private sectors of the space industries, with organisations such as NASA, ESA, Virgin Galactic, SpaceX and Blue Origin. Other mainstream online press media sites, dedicated private pages on the space industry and even Wikipedia for some content difficult to access due to language barriers (such as examples from Russia and China) were also utilised.

This book has seven chapters.

Chapter 1 – Introducing Space Tourism – presents the current space tourism industry, highlighting some key points in the history of space exploration, defining the typology of space tourism, explaining the status of commercial space tourism companies and describing the bodily experience of a space tourist.

Chapter 2 – Background to Sustainability – provides an insight into the discourse of global sustainability by defining the elements of sustainable development and global environmentalism, highlighting the discourse of climate change and factors enhancing it. It also presents a case study of a remote environment in relation to space.

Chapter 3 – Futures Forecasting – explores the process of reading weak signals, introduces new futures research models to assist in planning for a sustainable future, the Futures Map and the Sustainable Future Planning Framework, and investigates a future scenario of a space colony beyond the biosphere.

Chapter 4 – Planning Sustainable Space Tourism – explores the elements of sustainable planning, using destination life cycle models and linking space tourism and sustainable development on five different levels.

Chapter 5 – Space Tourism and Society – assesses some changes experienced during the human era of the Anthropocene, highlights social sustainability issues related to social fairness, space ethics and corporate responsibility, and examines global spaceports in terms of local impacts, using the United Kingdom as a case study.

Chapter 6 – Economics and Space Legislation – investigates the new space economy and space reusability, examines issues related to economic forecasts, market demand and pricing of space tourism, and reviews some considerations related to existing space regulations.

Chapter 7 – Visions of Sustainable Space Tourism – provides insights into different space tourism futures with visions gathered from various interviews, a professional Delphi panel and a public survey.

Finally, the concluding words highlight the major topics raised across the chapters and offer some last thoughts on future space tourism.

Your sustainable space journey is about to begin. Please enjoy it!

1 Introducing Space Tourism

> We are entering a new space age and I hope this will create a new unity. Space exploration has already been a great unifier, we seem able to cooperate between nations in space in a way we can only envy on Earth.
> Stephen Hawking, 2016

Introduction

By our very nature, humans are exploratory creatures and the Earth is no longer an adventurous enough place for some experienced tourists. Space tourism could be considered a new sector of adventure tourism, providing a novel opportunity to experience the extreme unknown (Toivonen, 2017). Expanding tourism into outer space could also prove to be a hugely significant endeavour for humanity, as travelling beyond the biosphere may fundamentally alter how we view ourselves, our place in the universe and our relationship to the Earth (Cohen & Spector, 2019). However, while space tourism has appeared within the context of adventure tourism and extreme sports in the tourism literature, these classifications are slightly problematic: actual space travel offers the tourist very few opportunities for expressing mastery over oneself or the environment. The space tourist is primarily a passive traveller, and all 'mastery' is left to professionals doing all the essential work required to make the space flight a success (Spector, 2020).

This chapter explores the history of space tourism, highlighting momentous historical milestones and governmental roles in the development of space tourism, followed by insights into the new space economy. It will then explore different space tourism activities, and introduce the main companies currently competing for the status of becoming the world's first private space tourism operator. Finally, it will consider safety aspects and the implications for the human body of visiting a space environment.

Historical Milestones

That's one small step for (a) man, one giant leap for mankind.
Neil Armstrong on his first step onto the Moon on 20 July 1969

The first century of aviation saw the industry move from the Wright brothers' miniature 'Flyer' to the development of passenger rockets to the Moon. However, space travel remains very different from aeroplane aviation, which has continued to be a growing and privately operated global industry though, until the 2010s, mainly a government monopoly (Finnair, 2008). Even though the 2010s saw the emergence of a new space economy combining private start-ups and entrepreneurs with traditional governmental actors, the argument that private enterprise has only factored prominently in space exploration in recent years has been opposed by MacDonald (2017) who, in his book *The Long Space Age*, demonstrated that private companies have actually always played a prominent role. Private space exploration, not including actual travel, can be traced back to the 19th and early 20th centuries when dozens of astronomical observatories were privately funded in America, with comparable relative economic significance to modern robotic spacecraft (MacDonald, 2017).

Occasionally, the feasibility of potential commercial space tourism has been under investigation, especially in research connected to NASA, showing enormous potential demand, at least based on people's desires to experience the Hollywood movie-styled setting of space. Over the past six decades, since the Soviet Union started the space race in 1961, there have been many visions about what space travel and tourism would entail, and what kind of conceptual designs of passenger space vehicles and infrastructure would be available.

During the Second World War, Germany's rocket programme proved the most significant transformative force for developing space technology. The first successful space flight was achieved in Germany on 3 October 1942, the first step on the path to suborbital passenger space flights (Launius, 2019).

Do you realize what we accomplished today? Today the spaceship was born. This third day of October 1942 is the first new era in transportation, that of space travel. (Walter Dornberger, 1942, following the first successful space flight on V2)

Instead of this possible new transportation scenario, rocket development was dominated by Cold War competition between the United States and the Soviet Union, which led to the production of tens of thousands of long-range missiles, resulting in a more than half-century delay in developing passenger space travel. In this light, the rockets used to launch satellites today, rather than being considered 'futuristic', could reasonably be described as 'obsolescent', as they could have been replaced by reusable launch vehicles several decades ago if policymakers had so chosen (Cole, 2015).

The US space programme emerged in large part because conquering space represented the ultimate symbolic power during the Cold War, when the United States and the Soviet Union were fearful of each other's capabilities and intentions (Launius, 2019). The Soviet Union's space programme declared a victory by successfully launching the first human being, Yuri Gagarin, into space on 12 April 1961. Orbiting the Earth, his flight lasted 108 minutes on the Soviet Union's Vostok spacecraft and, following his safe return to Earth, Gagarin became a cultural hero in the Soviet Union (Redd, 2018).

This victory motivated the United States to adopt the attitude of 'saving' the planet from 'evil intentions'. In his speech in May 1961, President Kennedy famously declared, 'If we are to win the battle that is going on around the world between freedom and tyranny, if we are to win the battle for men's minds... I believe that this nation should commit itself to achieving the goal, before this decade is out, of landing a man on the Moon and returning him safely to Earth'. The national hero attitude towards the Soviet Union's first cosmonaut was also reflected on early US astronauts, which helped NASA to accomplish bold future space plans with large budgets and start the race to the Moon (Ashford, 2002).

Project Apollo, which ran from 1961 to 1972 at a cost of $153 billion (in 2019 currency equivalent), presented the greatest engineering achievement of all time with the goal of the first human lunar landing on the Moon (Apollo, 2019). On 20 July 1969, Neil Armstrong and Edwin 'Buzz' Aldrin became the first humans to set foot on the Moon's surface, with Armstrong's world-famous words leading the way forward for future human endeavours in space. However, the American public questioned the value and cost of undertaking further human expeditions to the Moon at a time when society was in crisis over the Vietnam War, race relations and urban problems (Launius, 2019). The last Apollo mission was completed in 1972, resulting in the world-famous picture of planet Earth, The Blue Marble, that later became a symbol of environmental movements.

The 1960s space race between the United States and the Soviet Union provided a great opportunity to start forming and transforming space transportation for public use as well; however, this prospect was completely overlooked, with consequences that still have an impact today. Satellites have been launched using ballistic missiles or similar developments, with the fundamental disadvantage that they cannot be reused, creating higher launch costs and more emissions into the environment. The United States has been the leading power for most of the existing 'space age' and has carried responsibility for further developments of launchers and feasible planning for space vehicles.

Fully reusable launchers were still widely considered both feasible and the next logical step in the 1960s, but were not advanced for a number of reasons, primarily short-term invested interests, budget pressures and the political environment creating a lack of desire to pursue and further advance such projects (Ashford, 2002). Orbital space flights traditionally used launchers with complex single-use components based on ballistic missile technology. The X-15 model (from 1968) was the only fully reusable vehicle to have been to space over many decades; it had the capability to reach space and the ability to land like a conventional aeroplane, using wings for lift. Despite design teams at large aerospace companies carrying out studies on reusable launch vehicles, it wasn't until 2018 that private sector space tourism operator SpaceX introduced its reusable space vehicle in action.

First Space Tourists

The desire to connect with what lies beyond the Earth can be traced back to the early planetary science of Galileo, Greek astronomy, Aboriginal dreaming stories about the stars and even Captain Cook's first voyage to the South Pacific, in which his actual goal was to observe the transit of Venus across the Sun (Cater, 2019). The private sector's role in generating interest in space became obvious in America during the 19th century when many space observatories were built and funded privately (MacDonald, 2017). Later, various Hollywood productions further fed the human desire to explore the unknown and, during the 1960s, Pan American World Airways opened its first reservation lists for their first passenger flight to the Moon. The list, with over 93,000 names, was only closed decades later, upon the airline's cash-strapped demise in 1991.

In 2001, American Dennis Tito became the world's first space tourist to travel on board a Russian Soyuz rocket to the International Space Station (ISS). His trip made the option of space travel real for millions of 'ordinary' people without experience as astronauts. Later in the decade, wealthy businessmen who were able to afford the $20–$35 million cost also took journeys: South African Mark Shuttleworth (2002), American Gregory Olsen (2005) and Hungarian Charles Simoyi, who took two trips in 2007 and 2009 (Cater, 2019). A valid question, however, is whether these pioneering space travellers can be called 'tourists', as they all underwent months of training to become temporary astronauts, as well as participating in some scientific experiments in the space environment.

Space Adventures has so far been the only private company to arrange for paying passengers to go into space, in conjunction with the Federal Space Agency of the Russian Federation and Rocket and Space Corporation Energia (Space Adventures, 2019). The publicised price for flights to the ISS aboard a Russian Soyuz spacecraft ranged from $20 to $40 million and seven space tourists made eight space flights between

2001 and 2009. Some space tourists signed contracts with third parties to conduct certain research activities while in orbit. Russia halted orbital space tourism in 2010 due to the increase in the ISS crew size, using the seats that would previously have been sold to paying space flight participants for expedition crews. Orbital tourist flights had been set to resume in 2015, but the expected date for an operational start may now be in 2021 (Wikipedia, 2019b).

BOX 1.1 NASA'S SPACE TOURISM

- During the 1970s and 1980s there were various proposals (e.g. Space Habitation Design Associates) for removable passenger module cabins that could fit into the shuttle's cargo bay, carrying up to 74 passengers into orbit for up to three days.
- In the early 1980s, NASA introduced the Space Flight Participant Programme.
- In 1984, Charles D. Walker became the first non-government astronaut to fly with his employer, McDonnell Douglas, paying a sum equivalent to $96,000 today for his flight.
- In 1985, Christa McAuliffe was chosen to become the first teacher in space. Unfortunately, the Challenger disaster ended in tragedy at launch and the programme was cancelled.
- In 2003, a programme attempting to launch a second journalist into space was cancelled because of the Columbia disaster.
- In 2018, NASA announced the opening of the ISS to new commercial opportunities and private astronauts.

(Source: NASA, 2019a, 2019b)

BOX 1.2 RUSSIA'S SPACE TOURISM

- The space industry suffered a lack of funds post-Perestroika.
- In 1990, Tokyo Broadcasting System paid one of its reporters, Toyohiro Akiyama, to fly to Mir as a crew member and broadcast daily from orbit. The estimated cost of his space flight was between $10 and $37 million. As the flight was paid for by Akiyama's employer, he can be considered the first business tourist in space.
- In 1991, British chemist Helen Sharman became the first Briton in space in Project Juno, onboard Soyuz TM-12 to Mir, returning to Earth onboard Soyuz TM-11.
- During the 1990s, commercial 'edge of space' MiG military Russian fighter jet flights began.

- In 2001, Dennis Tito became the first paying tourist in space, arranged by space tourism company Space Adventures. He joined the Russian Soyuz TM-32 mission that docked to the ISS for a week.
- In 2002, South African internet entrepreneur Mark Shuttleworth became the second private tourist in space, paying $20 million for his eight-day stay aboard the space station.
- In 2003, American Gregory Olsen became the third space tourist, paying $20 million for a 10-day Soyuz excursion. He participated in some experiments on the ISS, hence preferring to be called a 'space flight participant' rather than a space tourist.
- Between 2001 and 2019, a total of seven tourists took private space tourism flights.
- In 2021, Russia plans to fly two space tourists to the ISS on a Soyuz spacecraft.

(Source: Fingas, 2019; NASA, 2019a)

New Space Economy

You want to wake up in the morning and think the future is going to be great – and that's what being a spacefaring civilization is all about. It is about believing in the future and thinking that the future will be better than the past. And I can't think of anything more exciting than going out there and being among the stars.
Elon Musk, 2019

The beginning of the millennium brought to light an interest in private space tourism start-up companies, forming an active and growing space tourism movement. The start of the commercial space tourism sector had already been forecasted in professional journals (Cornog, 1956; Dornberger, 1956) in the late 1950s, with the general assumption that ordinary people could also afford space travel after it had reached the same level of maturity as the airline industry, and that high-speed air transport would be available. However, until the 2000s, each Space Shuttle flight cost between $500 million and $1.5 billion, which is over 10,000 times more than, for example, the cost of a long-distance flight on a Boeing 747 (Ashford, 2002).

Using space for point-to-point transport across Earth is still not a viable near future possibility (Webber, 2010). For example, Virgin's existing spacecraft uses all its fuel to reach space and then glides back to Earth, a technique that would not work for a journey, for instance, from London to Sydney. However, there are progressive plans within the industry, such as at SpaceX, for point-to-point flights on rockets that could fly as many as 100 people around the world in minutes. The New York to Shanghai route would only take 39 minutes, rather than the 15 hours it currently takes by airplane (Sheetz, 2019a). In August 2020

Virgin Galactic announced an agreement with Rolls-Royce – a company with a history of making aircraft engines, including former Concorde aircraft – to develop and build an aircraft for supersonic air travel. Virgin Galactic provided a preview of its supersonic aircraft's design, after completing a review with NASA. The design would be able to hit Mach 3 (three times the speed of sound), using a Delta-wing aircraft, capable of carrying between 9-19 passengers and cruising at an altitude above 18 kilometres (Sheetz, 2020). With a top speed of around 3,700kmh, it could fly from London to Sydney in just five hours.

The 'new space' activities combine outer space and capitalism and are currently dictated by for-profit businesses owned by powerful and enthusiastic entrepreneurs engaging in different commercial space activities, such as SpaceX shooting a Tesla Roadster into outer space on board a Falcon Heavy rocket (Spector & Higham, 2019b). In the light of the space tourism start-ups, NASA started a $4.5 billion programme, Space Launch Initiative, to create the technology for a reusable launch vehicle. The XPRIZE competition was also set up to encourage entrepreneurial developments in space tourism. It offered a $10 million prize for the first non-governmental organisation to demonstrate a reusable and piloted suborbital vehicle capable of carrying three people to 100 kilometres and make two 100-kilometre flights within two weeks (Cater, 2019).

BOX 1.3   SOME BENEFITS OF THE LARGER-SCALE SPACE TOURISM INDUSTRY

- Existing space applications, such as satellites, will become less expensive and more widely available due to cooperative schemes.
- Likely to lead to a new golden age of astronomy and Earth science, resulting in more environmental knowledge and protection of the Earth.
- New local jobs and global economic expansion.
- The sharing economy benefits from 'new space'.
- New skilled jobs in specialised niches such as solar power manufacturing and handicrafts from space minerals.
- Introduction of hydrogen fuel for air and ground transport.
- The use of hydrogen may be safer than kerosene, increasing the safety factors for both space and air/ground travel.
- New living spaces and educational opportunities for the next generations.

(Source: Ashford, 2002; Toivonen, 2017)

In 2009, Ashford introduced an aviation approach strategy to space transportation, published in *The Aeronautical Journal*, for developing the first orbital spaceplane at low cost and risk, replacing the missile paradigms. His findings included the potential benefits of preliminary sizing and cost estimates of a simple lunar base and the cost of developing spaceplanes and other vehicles required. The total cost was 10 times less than that of the plans using large expendable launch vehicles (Ashford, 2009). The research came to light in a period when private sector space start-up companies were pursuing new approaches to transform the style of space flights, especially encouraging public–private partnership. NASA and other space agencies, including Chinese and Indian agencies, are currently involved in a new exploration of the Moon, comparable to the use of Saturn 5 in the US lunar programme in the late 1960s, to create a lunar base, assisting onward journeys to other planets, such as Mars.

Earlier, many directions had been predicted for the future of space tourism vehicle transportation, with rapid technological developments enhancing the step-by-step development sequences. Ashford's (2002) aeroplane approach to the spaceplane development strategy introduced four stages of development. The first suborbital space vehicles were forecast to perform similar to the X-15 model, and be able to carry test instruments, scientific experiments and, after completing safety and reliability checks, even some passengers. The first commercial space tourists would be able to enjoy the 'space experience', meaning feel the sensation of weightlessness for some minutes, see the curvature of the Earth and witness bright stars even during the daytime.

The second-stage space vehicles were forecast to use technology advanced from the first stage, with the ability to carry space station crews and mechanics to service satellites, and pioneer orbital tourism flights. The third stage would involve vehicles designed as small orbital spaceplanes, able to carry a five-tonne satellite or 50 people, for example, to space hotels. They would be capable of several flights into space per day and have a long service life as well as low maintenance costs. The fourth stage would include 'mature space planes', using technology currently being researched, to allow the use of a single stage, making a thousand-fold reduction in the cost of sending people to space and thus creating the market for mass tourism (Ashford, 2002).

Terrestrial Space Tourism

Terrestrial space tourism is quite well established and includes Earth-based simulations and entertainment experiences such as visits to space observatories, museums and exhibitions, and stargazing with a telescope, in places such as Manu Kea Observatory in Hawaii. The Kennedy Space Visitor Center in Florida is the most popular terrestrial space tourism facility, hosting over 1.5 million visitors per year despite also being an active spaceport. Other forms of terrestrial space tourism include cyberspace tourism experiences, such as virtual gaming environments and virtual reality travel (Cater, 2019). Space movie-related enhanced travel, eclipse tours, meteorite rock collecting, UFO hunting and seeing the Northern Lights are also examples of Earth-bound space tourism activities. As these activities do not require participants to physically leave planet Earth, terrestrial space tourism already serves as a good example of sustainable practice in space tourism.

Virtual space tourism

Virtual reality 'enables one to gain an experience of the outer space environment without leaving the Earth. It democratises participation in future space tourism and expands its itineraries while also altering the practices of tourism' (Damjanov & Crouch, 2019: 117). For many people, space travel represents one of their biggest dreams, and the first space tourism products may be based on this curiosity factor. For decades, space-related sci-fi television shows such as *Star Trek* and Hollywood films such as *Star Wars* and *Ad Astra* have produced fascinating and addictive stories about space travel for many different generations.

Future virtual reality gadgets may offer many advantages compared to real space travel, as they are more environmentally friendly, safer, cheaper and an easily accessible opportunity to discover space for those with existing medical conditions. There are already public virtual reality tours available in various locations, such as in NASA's Kennedy Space Visitor Center in the United States, which presents realistic 3D simulations of the soil of Mars (Kennedy Space, 2019). When commercial space tourism takes off, some people will be unable to participate for financial or health reasons, creating a market for virtual reality platforms (Beck, 2016).

In 2015, start-up company SpaceVR began to produce 360-degree content to make space travel accessible to everyone, allowing the user to experience footage from a short journey into space and back to Earth. Future plans involve sending cameras with the ability to shoot 3D in 360-degrees to the ISS and to a specially designed satellite, equipped with two 240 field of view (FOV) 4K sensors, to capture high-resolution footage from all around the satellite. The data will be relayed back to Earth through X band radio. The footage will be run through a post-production process and finally converted into interactive 360-degree videos that work on all major head-mounted displays (HMDs) such as Samsung Gear VR and Oculus Rift (SpaceVR, 2019).

Stargazing and the Northern Lights

Night-time stargazing and watching Aurora (the Northern Lights) allow people to explore the universe from Earth, especially in areas without significant light pollution. Some features can be witnessed with the naked eye, while others can only be seen with the assistance of a telescope. A tourism industry already exists offering stargazing packages around the world, especially during solar eclipses and in the northern regions such as Lapland in Scandinavia, where packages to 'witness the Northern Lights' are marketed to tourists worldwide (HeliosTour, 2020).

Educational games

Space exploration tourism can also be accomplished via interactive media and contemporary video games which are 'a major site at which future visions of space tourism can be displayed and directly interacted with, allowing players to experiment with modalities of extraplanetary transit' (Ceuterick & Johnson, 2019: 105). For example, a Finnish start-up company, Space Nation, captured worldwide attention in 2017, claiming to build a social media-based app that would eventually produce a reality TV show and a grand winner with a ticket to space.

More traditionally, family board games can offer sustainable space exploration. One example is 'Crash Landing', with the concept of the game based on surviving a crash landing 60 kilometres away from the nearest lunar base. The idea is to think about the environment on the Moon and try to determine which resources would be of most help. Other ways to enhance interest in space include visual stories narrated by real astronauts on DVDs, online games such as National Geographic Kids Passport to Space, NASA's Passport to Explore Space and online scientific tools which allow users to create a pocket solar system (Astro Society, 2019).

Atmospheric and Suborbital Tourism

According to Webber (2019: 165), 'the key to understanding the whole field of space tourism is reusability. The potential market of space tourists is large enough that economies of scale make sense. The price elasticity of demand for human payload is so high that this produces the need for reusable, therefore lower-cost space access and hence the need for reusable rockets and spacecraft'. So far, such space tourism has been practised mainly at atmospheric altitudes, with an existing military-designed fleet. However, the development of space capsules for low-Earth orbit in the commercial sector in the near future are mainly designed to be reusable.

Zero gravity flights

Zero gravity flight is an experience at a higher altitude, in which the passenger does not leave Earth but has the opportunity to experience true weightlessness. For example, the ZERO-G modified Boeing 727-200 performs parabolic arcs to create a weightless environment allowing passengers to float, flip and soar as they would in space. The cost is about $6000 per person and flights reach an altitude of 32,000 feet (Space Adventures, 2019). Jet flights of up to 20 kilometres have also been offered with Russian MiG crafts. Zero gravity flights are an excellent opportunity for passengers to experience and familiarise themselves with some of the elements of space, even the possible feeling of motion sickness caused by the High-G acceleration phase of the flight.

Edge of space flights

The edge of space flight takes place on the upper edge of the Earth's lower atmosphere and represents the altitude limits of jet aircrafts. Russian MiG flights, offered by various private entrepreneurs, have provided space tourism opportunities since the 1990s, at a cost of around $20,000. Passengers need to undergo high gravity training as the jet pilot can take them up to 80,000 feet, where the plane seems to be floating, providing a view of the blackness of space, the curvature of the Earth and the atmosphere that looks like a blue fog (MiGFlug, 2019).

Suborbital flights

A suborbital space flight occurs when a spacecraft reaches outer space, such as the Karman line at 100 kilometres, but its trajectory intersects the atmosphere or surface of the gravitating body from which it was launched so that it will not complete an orbital revolution. The journey combines the excitement of a rocket-assisted jet flight to orbit with an extended period of weightlessness and a view of Earth (Anderson, 2005).

Suborbital tourism flights focus on attaining the altitude required to qualify as reaching space. SpaceShipOne won the Ansari XPRIZE competition in 2004, which invited commercial companies to compete to be the first in space, with the requirement of completing two successful flights within a two-week period. On 13 December 2018, Virgin Galactic's VSS Unity achieved the first suborbital flight status, reaching an altitude of 82.9 kilometres, officially entering outer space by US standards. In February 2019, for the first time, a passenger joined the team on board and sat and floated within the cabin during the flight (Virgin Galactic, 2019a).

Suborbital space tourism will be the first stage of so-called common space tourism as it does not require passengers to partake in astronaut training beforehand. Naturally, however, some intensive flight training is required, covering weightlessness, inflight acceleration and safety and equipment training. To date (August 2020), suborbital space tourism has not started, despite it being thought at the time of the Ansari XPRIZE competition that the suborbital tourism experience derived from the winning SpaceShipOne would be available to the public within four years – instead it has so far taken over 15 years, safety issues and regulations being the biggest contributors to the delay (Webber, 2019). However, the start of such space flights is just around the corner.

Polar orbital tourism

Currently, most orbital missions are launched from spaceports located relatively close to the equator, especially in the case of satellite launches benefiting from the east–west equatorial orbit as it allows them to be positioned to orbit the Earth at the equator in 24 hours. This makes them appear to remain in a fixed position in the sky. From the polar orbit, the objects orbit the Earth in a north–south direction and the polar orbit's 90-degree inclination provides more complex coverage of the Earth's surface. Polar spaceports make reaching such orbits more efficient, because all of the spacecraft's energy can be used for north–south velocity (Anderson, 2005).

In 2007, it was announced that Virgin Galactic and Spaceport Sweden in Kiruna, in the north of Sweden, are planning to begin a mutual operation to start offering space tourism, promoting the option to witness the Northern Lights from the 'inside' (STT, 2007). The Pacific Spaceport complex in Alaska may also present a unique opportunity in the future for space tourists, offering a chance to orbit the Earth over the North and South Poles and witness views such as polar ice caps, frozen seas, the Northern Lights and the Arctic mountains.

Orbital and Astrotourism

The International Space Station

The ISS is a space station in low-Earth orbit, structurally similar to an artificial satellite. The ISS is a joint project between five participating space agencies: NASA (United States), ESA (Europe), Roscosmos (Russia), JAXA (Japan) and CSA (Canada). Use of the space station is established by intergovernmental treaties and agreements and the station, which is reached by a Space Shuttle, serves as a microgravity and space environment research laboratory and also as a test site for equipment required for missions to the Moon and Mars (NASA, 2019c).

The station has so far been the ultimate destination for adventurous space tourists, offering a floating apartment complex with several activity areas and observation posts to take advantage of good views of Earth and outer space. While staying at the ISS, tourists have also been able to operate and visit different modules and devices developed by numerous countries (Anderson, 2005).

On 7 June 2019, NASA announced that it is to open the ISS to new commercial opportunities and private astronauts so that the industry, innovation and ingenuity of the United States can accelerate a thriving commercial economy in low-Earth orbit. NASA will enable private astronaut missions of up to 30 days to perform duties that fall into approved commercial and marketing activities, with the first mission planned for 2020. If supported by the market, the agency can accommodate up to two short-duration private astronaut missions per year, which will be privately funded, as well as dedicated commercial space flights (NASA, 2019f).

In January 2020, NASA announced that it had selected the first commercial destination module for the ISS. At least one habitable commercial module will be attached to the space station's Node 2 forward port by Axiom Space of Houston. It will demonstrate its ability to provide products and services and begin the transition to a sustainable low-Earth orbit economy (NASA, 2020b).

Private space stations

Bigelow Commercial Space Station is a private orbital space station operated by Bigelow Aerospace. On 8 April 2016, this inflatable module was attached to the ISS with the assistance of SpaceX where it has since been tested. The company has made space as a commercial destination a reality through the creation of lightweight modules and a habitable space environment at a lower cost. However, it is currently mainly used as cargo stowage (Bigelow Aerospace, 2019). In the future, space tourists visiting the private space station may be offered smaller living spaces, but as it is a private sector facility, access will most likely be granted more easily than to the ISS.

The Moon

In December 2017, US President Donald Trump announced that the United States will send astronauts back to the Moon and Policy Directive 1 was signed to make a change to provide an integrated programme for a human return to the Moon followed by missions to Mars and beyond that would be US led with private sector partners such as SpaceX. With the new policy, NASA (2019g) will 'lead an innovative and sustainable program of exploration with commercial and international partners to enable human expansion across the solar system and to bring back to Earth new knowledge and opportunities'.

> The directive I am signing today will refocus America's space program on human exploration and discovery. It marks a first step in returning American astronauts to the Moon for the first time since 1972, for long term exploration and use. This time we will not only plant our flag and leave our footprints – we will establish a foundation for an eventual mission to Mars, and perhaps someday, to many worlds beyond. (Donald Trump, 2017)

So far, the Moon is the only reachable physical space destination, with its low gravity, craters and landscapes. To date, only 24 people have visited the Moon, the last visit being in 1972 (SpaceX, 2019a). In 2017, SpaceX announced plans to fly passengers on a private trip around the Moon on a Dragon Spacecraft: 'Like the Apollo astronauts before them, these individuals will travel into space carrying the hopes and dreams of all humankind, driven by the universal spirit of exploration' SpaceX (2017).

However, this flight was later cancelled due to Falcon Heavy's inaugural launch. In 2018, it was announced, again from SpaceX, that they are planning to send Japanese billionaire Yasuku Maezawa to the Moon. He will be the company's first private customer to travel around the Moon on the Big Falcon Rocket in 2023. Yasuku Maezawa, who launched Japan's largest online retailer with a net worth of almost $3 billion, has already declared, 'I choose to go to the Moon with artists!'. He is planning to invite up to eight artists from around the world to join him. They will be asked to create a 'masterpiece' after returning to Earth, inspiring the 'dreamer within all of us' (Grush, 2018). Bigelow Aerospace envisions that a tourist-supported lunar station could be built for $2 billion, with staterooms providing 40% of the Earth's gravity, allowing visitors to jump vast distances, but to be shielded from dangerous radiation from space (Anderson, 2005).

Mars

There are many strategic, practical and scientific reasons for humans to explore Mars, and NASA's (2019d) mission statement for the current Mars Exploration Programme is 'to explore Mars and provide a continuous flow of scientific information and discovery through a carefully selected series of robotic orbiters, landers and mobile laboratories interconnected by a high-bandwidth Mars/Earth communications network'. From NASA's point of view, exploring Mars will improve the quality of life on Earth and, because in many aspects Mars is the most Earth-like of all the other planets in the solar system, it could someday be a destination where humankind could survive.

SpaceX has an interplanetary transport mission in which building bases on the Moon and cities on Mars will require affordable delivery of significant quantities of cargo and people. However, SpaceX's aspirational goal is already to send the first cargo mission to Mars in 2022. The objectives are to confirm water resources, identify hazards and put in place initial power, mining and life support infrastructure. The second mission, carrying both cargo and crew, is projected for 2024, with the primary objectives of building a propellant depot and preparing for crew flights in the future. The ultimate goal is to build a self-sustaining colony on Mars, which most likely could serve as an infrastructural base for space tourism as well (SpaceX, 2019a).

Doubts have been expressed as to whether the private companies championing human missions to Mars will actually succeed. Launius (2019: 49) claims: 'Firstly, Mars is a difficult undertaking for robotic probes, but especially human missions. Secondly, there is no compelling rationale at present for undertaking the mission rather than prestige and bragging rights, which is not a sustainable reason. Thirdly, the costs of such an endeavour may well be in the one-trillion-dollar range and no private sector will pursue this end without an enormous capability for return investment'.

Space Tourism Companies

Virgin Galactic

> Together we open space to change the world for good.
> Virgin Galactic, 2019a, motto

Virgin Galactic, founded in 2004, is a space flight company belonging to the Virgin Group owned by adventurer Sir Richard Branson. The company's main aim is to provide suborbital commercial space flights to space tourists and suborbital launches for space science missions. Its aim is also to transform the current cost, safety and environmental impact of space launches and pioneer the next generation of reusable space vehicles.

SpaceShipTwo, based on the Ansari XPRIZE-winning SpaceShipOne concept, is a rocket plane lifted initially by a carrier aircraft before independent launch. Originally the first space tourism flight was due in 2009, but there have been delays on a number of occasions, especially after the test flight of SpaceShipTwo, VSS Enterprise, went seriously wrong. The company had already sold tickets costing between $200,000 and $250,000 each for the estimated 90-minute flight for up to 700 passengers.

However, on 13 December 2018, VSS Unity achieved the first suborbital commercial space flight, followed by a successful flight with a team member in a passenger seat. However, this did not count as the first official space tourism flight as the passenger was not a paying customer, but astronaut trainer Beth Moses, who also validated some of the cabin design elements. On February 2020, Virgin Galactic successfully completed another vital step on its path to commercial service, relocating SpaceShipTwo, VSS Unity, to its commercial headquarters at Spaceport America. This captive carry flight provided an opportunity to evaluate VSS Unity for over three hours at high altitude and cold temperatures (Virgin Galactic, 2019a, 2019b, 2020). In July 2020 Virgin Galactic revealed the final design of the interior cabin of its VSS Unity spacecraft in a virtual event. The cabin was designed with customer experience chiefly in mind optimizing the safety and comfort with custom fit tailored seats (that also double as holdouts during the free float), 17 passenger accessible windows throughout the interior rounded with soft material and fitted with integrated cameras to capture photos for the space tourism experience documentation (Etherington, 2020).

SpaceX

> Launch and land and relaunch. Guys! It's so simple.
> (Un-official SpaceX motto used by Elon Musk)

SpaceX is a private aerospace manufacturer and space transportation company founded in 2002 by Tesla entrepreneur Elon Musk. The ultimate aim of the company is to colonise Mars and enable people to live on other planets as well as to revolutionise space technology and reduce space transportation costs. The fleet includes the Falcon launch vehicle and Dragon spacecraft families. In 2008, SpaceX was the first private space company to successfully launch and orbit, to recover a spacecraft in 2010; in 2012, it was the first private company to deliver cargo to and from the ISS; and in 2017, it accomplished the first reuse of an orbital rocket with Falcon 9 (SpaceX, 2017, 2019b, 2020).

SpaceX is working in partnership with NASA and has delivered cargo and transported astronauts to the ISS. SpaceX has a reusable launch system development programme with success in using the drone ship landing platform. In 2016, Elon Musk unveiled SpaceX's Interplanetary Transport System, which is a privately funded initiative to develop space flight technology for use in crewed interplanetary space flight. In addition, an updated configuration named 'Starship SN2' has been scheduled for 2021 and it is planned to be fully reusable from the beginning (SpaceX, 2019a, 2019b). In late 2021, SpaceX plans to send three tourists on a 10-day trip to the ISS using its Falcon 9 rocket and its new Crew Dragon spacecraft. The orbital vacation is part of a deal that SpaceX signed with Houston-based start-up Axiom Space and this mission will be the first fully private trip to the ISS (O'Kane, 2020). On 30 May 2020 SpaceX launched two NASA astronauts, Robert Behnken and Douglas Hurley, toward the International Space Station on Falcon 9 rocket, marking the first time in the history that a commercial built and operated American crew aircraft carried humans into Earth's orbit (NASA, 2020a).

Blue Origin

> Gradatim, Ferociter (Step by Step, Ferociously)
> Blue Origin, 2019, motto

Blue Origin was founded in 2000 by one of the world's richest men, Jeff Bezos, and has been developing technologies to enable private human access to space with the goal of dramatically lowering costs and increasing reliability. Blue Origin claims not to be in a space tourism race, promoting a slower step-by-step development process to its rival companies and traveling to space specially to benefit the Earth. The company's mission is 'to build a road to space with reusable launch vehicles so our children can build their future' (Blue Origin, 2019).

Blue Origin has been developing several technologies, with a focus on rocket-powered vertical take-off and landing vehicles to access suborbital and orbital space. The company has also been contracted to work for NASA under several development efforts. Blue Origin started developing systems for orbital human spacecraft prior to 2012, but the space vehicles are still undergoing testing (New Shepard and New Glenn). However, since 2015, the uncrewed space vehicles have already reached beyond the Karman line, with successful landings.

In 2019, Jeff Bezos unveiled plans for a flexible Moon lander, 'Blue Moon', to deliver a wide variety of small, medium and large payloads to the lunar surface with soft landings to enable a sustained human presence on the Moon, set to be ready by 2024. On April 30 2020 the Blue Origin National Team was selected by NASA to return the humans to the Moon and began to develop the Artemis Human Landing System with an intention to return to the lunar surface permanently (Blue Origin, 2020). This would complete the dream of an 18-year-old Bezos, describing in an interview with the *Miami Herald* in 1982 that, 'I am to build space hotels, amusement parks and colonies for two million or three million people who would be in the orbit. The whole idea is to preserve the Earth and the goal to be able to evacuate humans so the planet would become a park' (Whoriskey, 2013) (Blue Origin, 2019; Wikipedia, 2019a).

CosmoCourse

> We are talking about sub-orbital tourist traffic. The launch vehicle, the descent vehicle, and the engine are currently being developed.
> Sergei Zhukov, Co-Leader of CosmoCourse Project, RT, 2019

In 2017, private Russian company CosmoCourse received a licence for space activities and is planning to create a reusable spacecraft for touring in space. The company plans to offer tourists an excursion close to the height of the Karman Line (100 km) on an open trajectory before descending by parachute or engine powered aircraft. The permission to participate would be granted subject to three days of training and a medical examination. The ticket price will be around $250,000 and the flight time for tourists will be 15 minutes, including 5 minutes in a state of weightlessness while freely floating around the cabin looking at the Earth (Dorozhkina, 2017).

High-altitude balloon companies

In 2020, Space Perspective (launched by the founders of near-space exploration company World View Enterprises) started to promote its future high-altitude space flights in a balloon-borne pressurized cabin, complete with a bar and restroom and communication capabilities with friends and family below. The estimated US$125,000 cost of each flight would carry eight passengers at a time to the edge of space and experience a six-hour flight. The passenger cabin, lifted by a huge hydrogen-filled balloon, would climb at a sedate 20 kilometres per hour to an altitude of about 30 kilometres, followed by a slow descent to splash down in the Atlantic Ocean where a recovery ship would be standing by to secure the cabin and crew. 'Spaceship Neptune' will start operating from leased facilities at NASA's Kennedy Space Centre (Florida) within the next four years, with piloted test flights starting in 2021 (Space Perspective, 2020).

For the last 10 years, a Spanish company, Zero 2 Infinity, has also been testing high-altitude balloons, launching small payloads on the Bloostar orbital launcher to high altitudes to test low-Earth orbit satellites and commercial products. The company plans to extend its services to commercial tourism flights in the future, already highlighting that their launch system will have a significantly lower impact on the environment compared to the other options and also that access to space should be affordable and secure for everyone (Zero 2 Infinity, 2020).

Space Adventures

Space Adventures has organised flights for the world's first private space explorers, such as Dennis Tito and Mark Shuttleworth. The company offers a variety of space-related programmes, including zero gravity flights, space flight missions to the ISS or around the Moon, record-breaking orbital missions and various training and space flight qualification programmes. In 2019 Space Adventures announced that they had signed a contract with Roscosmos (which is a state corporation responsible for the wide range and types of space flights and cosmonautics programs for the Russian Federation) to fly two paying customers and a single professional cosmonaut in a special Soyuz in late 2021, Energia Corporation (the Soyuz constructors) also confirming that a modified version of the current Soyuz MS model is being developed (Quine, 2020). In February 2020, Space Adventures announced an agreement with SpaceX to launch private citizens on the Crew Dragon spacecraft, which is used by SpaceX to transport NASA astronauts to the ISS. This will provide up to four individuals with the opportunity to break the world altitude record for private citizen space flight. The flight will be capable of reaching twice the altitude attained by any prior civilian mission or space station visitor (Space Adventures, 2020). According to Quine (2020), 'the fact that Space Adventures is dealing with both Roscosmos and SpaceX is interesting, and doubtless reflects a desire to drive up the supply of seats, drive down the price, and develop a level of competitiveness which has not previously been possible in this sector'.

Orbital hotels

Currently, the only existing habitable human-made orbital infrastructure in space is the ISS, which requires a very high standard of engineering because of its older forms of technology and the great cost of repairs in case of a fault. In the future, orbital hotels for tourists could, for example, be located about 500 kilometres from the Earth, in orbit, with one rotation around the Earth taking approximately nine hours. If a tourist spent four days in an orbital hotel, they would orbit the Earth in a weightless environment about 10 times.

The facilities of such orbital hotels would include restaurants, relaxation and well-being areas including a gym, plant rooms and an auditorium for business gatherings. All spaces would be surrounded by large windows facing the Earth and, for more detailed stargazing, there would be access to telescopes and maps with comprehensive descriptions of the area. They would have regular contact shuttles from the Earth, with a flight time of about 20 minutes, and the hotel would be especially popular with honeymooners (Finnair, 2008).

Aurora space hotel

> Our long-term vision is to sell actual space in those new models. We're calling that a space condo. So, either for living or sub-leasing, that's the future vision here – to create a long term, sustainable human habitation in low Earth orbit.
> Frank Bunger, CEO, Orion Span, 2018

Orion Span is planning to open a luxury hotel in orbit 200 miles above the Earth in 2021. The single-module commercial space station will start accommodating guests in 2022 and will offer private suites with numerous windows. A 12-day visit to the hotel will cost from $9.5 million and guests will need to attend three months of pre-flight training before their hotel stay. The size of the Aurora station will be similar to a private jet cabin, with hotel personnel planned to be former astronauts (RT, 2018).

Becoming a Space Tourist

According to Dickens (2019), Henri Lefebre's conceptualisation of society's spatialisation could be used to understand society's relations to outer space. Outer space can be viewed as a new spatial fix, as capitalism always requires new 'outsides' from which to obtain extra resources and ensure continued growth. However, the dynamic could be seen in a more negative than positive light, as there is a high dependency on machinery. For example, the representation of astronauts in their spacesuits as half human and half machine replicates the hybrid trans-human, as the human body cannot cope alone in the space environment (Dickens, 2019: 209).

Safety

Safety is the most crucial issue in ensuring future passenger space vehicle operations. Ashford (2002) estimated that crewed space flight has been approximately 10,000 times more risky than airline flight, a calculation made from statistics showing that by the start of the millennium the United States had made 137 crewed space flights, with one fatal accident (the Challenger Space Shuttle crash in 1986) and the Soviet Union/Russia had two fatal accidents in 94 crewed space flights, making the number of flights per fatal accident around 100. In 2014 the VSS Enterprise, a Space-ShipTwo experimental spaceflight test vehicle operated by Virgin Galactic crashed in the Mojave Desert in California, killing one pilot and injuring the other. The accident was caused by a combination of human error (the co-pilot unlocking the braking system too early) and inadequate safety procedures, explored in an inquiry concluded in a nine-month investigation by The National Transportation Safety Board (Gajanan, 2015).

It was speculated that the fatal explosion would lead to more industry regulations and there were fears about the survival of the space tourism industry given the safety image damage caused by such a crash (Vergano, 2014). To meet operational safety standards, the commercial space tourism companies must participate in and pass rigorous flight-testing programmes by the Federal Aviation Administration (FAA) in the United States or the European Aviation Safety Agency (EASA) in Europe, before they can be issued with a passenger-carrying certificate (Ashford, 2002).

In the early 1990s, research into the prospects for space tourism, looking at reasons why people were not interested in travelling to space, was presented to the European Aerospace Congress (Abitzsch, 1996). A breakdown of the main reasons was that 30% of the respondents cited safety concerns and 10% thought the idea of space tourism simply wasn't realistic. However, it was pointed out that, if space tourism became operational and its safety proven, these two groups would most likely change their minds, adding to potential demand for space tourism. The study also concluded that the desire to travel to space is quite consistent in all cultures, encouraging general prospects for space tourism, which should be conducted as a global enterprise with a potential demand assured even after the early expensive phase (Abitzsch, 1996).

BOX 1.4 REASONS WHY CREWED SPACE FLIGHTS HAVE BEEN RISKIER THAN AVIATION

- The throwaway launchers using ballistic missile technology: previously it has not been practical to aim for high safety with expendable vehicles because their high cost per flight precludes a full flight-testing programme.
- The production line of space vehicles has traditionally not been able to perform inflight before actually being used as the vehicles are only flown once. This practice has been completely opposite to that of the airline industry, in which many test flights are made by the manufacturer before operational passenger flights.
- There are increased hazards for high-speed rocket flights such as the use of high-energy rocket propellants, re-entry heating and stability control at high speeds.
- Safety challenges due to the space environment, such as the effects of loss of cabin pressure, space radiation and space debris.
- A long duration in space has been proven to have a negative impact on the human body.
- In case of an emergency, the evacuation and aftercare in space are extremely challenging to operate from the Earth.

(Ashford, 2002)

Medical considerations

If one were to imagine a future in which space travel has become the norm and is available to the masses, almost equalling the situation in aviation today, different preparations would need to be taken into account.

Firstly, there are several medical dilemmas, which also reflect ethical considerations, in space tourism. The medical implications are mostly associated with space flights and the ethical problems surround the safety of space travel. Unlike professional astronauts, who have undergone extreme health tests and are under constant monitoring, the space tourist will most likely be required to undergo a doctor's physical check-up and sign an insurance form to exclude the space tourism company from liability in case of a sudden injury or death. It is also possible that space tourism businesses may face liability for passengers' deaths or injuries, if in the case of such events they are found to have been negligent. The company's duty to disclose potential passenger risks could even include issues such as space sickness, loss of bone density and risks associated with microgravity (Marsh, 2006).

Some former NASA astronauts have given advice for first-time spaceflight participants emphasising that the human body will experience a range of new sensations during the space flight. To be able to gain the best possible tourism experience from one's short duration in space, both the physical and mental well-being of a space tourist are a priority, hence a recommendation to prepare for the change to a weightlessness environment by taking a zero-G flight or scuba diving in advance (Waldek, 2020).

The space environment

Space cannot be accessed without the proper protection to shield the traveller from space radiation and the temperature, and pressurised oxygen must be provided. If the body does not get enough oxygen, there is a risk of the person becoming unconscious and dying from hypoxia. The space flight must provide suitable pressurised conditions to avoid such dangerous effects on the human body.

There is a serious health risk from cosmic rays as space does not have the Earth's protection, exposing tourists to high levels of radiation. Health studies supported by NASA have demonstrated that radiation may harm the brain and, for example, lead to the onset of Alzheimer's disease (NASA, 2019b). High levels of radiation also damage the lymphocytes and weaken the immune system. In the event of a solar flare, a fatal radiation dose could be received within minutes. However, the Earth's magnetic field partly protects the surrounding area in space, such as the area where the ISS is located, meaning that suborbital space flights will benefit from this shield as well (NASA, 2020c).

The human body will be much more vulnerable to the risk of radiation in potential missions to Mars, as the radiation risk is long-lasting, impacting, for example, the body's ability to protect itself against diseases. In addition, the longer journey time exposes travellers to the possibility of experiencing an unexpected solar storm (Gueguinou *et al.*, 2009).

Effects on the human body

NASA has reported health issues, stating that entering space has some negative impacts on the human body. In 2019, NASA performed an 'Astronaut Twin Study' in which one twin spent a year in space on the ISS while the other twin stayed on Earth (Garrett-Bakeman, 2019). Several long-lasting changes were reported in the twin in space compared to the twin who remained on Earth, including alterations to DNA and cognition. These kinds of scientific findings are important in determining the impact of long-term space travel on the human body, for example on the human mission to Mars currently planned by SpaceX (NASA, 2019b).

Previous studies on astronauts have also demonstrated several physical effects due to long-term weightlessness, including muscle atrophy and deterioration of the skeleton (Kanas & Manzey, 2008). Other significant effects have included decreased production of red blood cells, balance disorders, slowing of cardiovascular system functions, eyesight disorders and changes in the immune systems. Minor health effects have been loss of body mass, nasal congestion, sleep disturbance and fluid redistribution, causing a moon-face appearance (Neergard & Birenstein, 2019).

Physiological effects

The conditions on board a space flight or in space are very different to those to which humans are normally accustomed. In particular, longer-term 'living' in the space environment has an impact on the body in different ways, causing loss of proprioception, changes in fluid distribution and deterioration of the musculoskeletal system. Even though space vehicles and spacesuits are technologically designed to provide the necessary support of breathable air, research on astronauts taking longer trips to space has presented significant changes in the position and structure of the brain. The longer an astronaut had spent in space, the more brain changes had caused them to age prematurely (Griffin, 2018).

G-force, weightlessness and motion sickness

During ascent and re-entry, gravity is experienced at several times the normal level on Earth. If the person is untrained, the usual limit is 3G, thereafter blackout may occur, especially in the vertical direction, causing blood to flow away from the brain and the eyes (Voshell, 2004). For the purpose of future tourism ventures, the spacecraft needs firstly to be designed to maintain G-forces within comfortable limits.

Exposure to weightlessness has been reported to affect astronauts' health as the physical conditions on Earth differ from the surface of the Earth. Short-term exposure can cause space adaptation syndrome, which is a self-limiting nausea caused by derangement of the vestibular system. Long-term exposures can contribute to loss of bone and muscle mass, increasing the risk of a general injury (ESA, 2020). Motion sickness is one of the most common problems experienced by astronauts in the initial hours of weightlessness, causing nausea, headache and vomiting. Motion sickness has been reported to last up to three days of the journey, after which time the body usually adjusts to the new environment (Kanas & Manzey, 2008). As with sea sickness, medication is available to ease the symptoms, but vomiting into a spacesuit can be fatal to space tourists.

Vision, taste, stress, sleep

NASA research on animals such as monkeys has reported that there is a risk that eyesight may become blurry after a longer period in space (NASA, 2019b). As many as half of the astronauts have reported blurred eyesight in both near and distance vision during their missions, with, in some cases, recovery not occurring until years after the mission. It has also been reported that there are changes to the sense of taste and a favourite food or drink may taste different in space. Flavours need to be very strong to combat the loss of taste (Newsela, 2016).

Human nature is one of the biggest threats to the success of space travel, as the environment itself presents participants with many stressful factors. A group of virtual strangers with different cultures, languages and morals enclosed in a small space could quickly lead to issues even on the ground, let alone in space. The longer the duration of the journey, the more the feeling of isolation, especially from friends and family, and the frustration of missing out on important events, such as the birth of a child, may occur. The space environment also impacts on sleep due to the poor illumination during the daytime hours and the light and dark cycles in the spacecraft. This can cause increased fatigue, decreasing daytime productivity. In the long term, disturbed sleep may lead to impaired focus and depression (Marsh, 2006).

Conclusion

Sir Richard Branson (founder of Virgin Galactic) has defined the significance of the momentum well when he says, 'We are at the vanguard of a new industry determined to pioneer twenty-first century spacecraft, which will open space to everybody – and change the world for good'. As we review the history of space exploration, we can see that the formation of a commercial space industry will be relevant in terms of not only the future of transportation, but also for the future imagination of society. Since space exploration began, it has greatly expanded the horizons of human knowledge and activity. In terms of significance, the entry of the human race into outer space may be compared with such landmarks in the development of human culture as the discovery of fire and the wheel, paving the way for advancement to greater geographical distances and discoveries (Vereshchetin *et al.*, 1987).

The greatest unfulfilled desire of world-famous cosmologist, Professor Stephen Hawking, was to travel to space, but, sadly, commercial space tourism had not commenced before his death in 2018. However, despite his disability, Hawking managed to experience a zero gravity flight on a specially modified Boeing 727-200, withstand the G-forces involved in space flight and show the world that people with disabilities can also become future space travellers (BBC, 2007). In the future, the incipient space tourism industry is likely to offer a variety of space-related adventures, from suborbital jumps to holidaying in orbital hotels, and even landing on the Moon or Mars. Rapidly advancing reusable technology will open up many further possibilities for all kinds of space exploration with far-reaching implications. Additionally, the experience of space tourism will not be limited to the authentic physical bodily experience, but will be possible through enhanced virtual reality gadgets and other means of terrestrial space exploration, making some version of space tourism accessible to everyone.

2 Background to Sustainability

> We are determined to ensure all human beings can enjoy prosperous and fulfilling lives and that economic, social and technological progress occurs in harmony with nature.
> UN, 2019b: 5

Introduction

The world we live in can be very cruel not only to the environment, but also to our social values and the systems based on them. According to the UN (2019b), we are living in a time of immense challenges to sustainable development, with climate change being one of the greatest problems of our time, its adverse impact undermining the ability of all countries to achieve sustainable development. The way in which we view global sustainability can be something of a jigsaw as it may be considered an ideology in its own right yet also include environmentalism and eco-colonialism. In hegemony, sustainability may be considered a contested concept involving globalisation, discourse itself assuming objectivity towards environmental issues and the intellectualisation of travel (Mowforth & Munt, 2015). However, the environment, natural or artificial, is the most fundamental ingredient of the tourism product in contemporary society. When new tourism activities, such as space tourism, start to occur, the environment modifies immediately to facilitate the new venture, creating a demand for different environmental considerations.

This chapter provides background to the mechanisms of sustainable development in general as well as insight into the discourse of sustainability. The aim of this exploration of the theoretical context is to characterise the features of the era in which space tourism will start operating. The examples provided in this chapter relate mainly to environmental sustainability, with a focus on defining the major factors impacting the world's future environmental carrying capacity. Society's increased global environmental awareness will also be explored, along with a case study that uses the remote Arctic as an example.

Defining Sustainable Development

Bruntland's (1987) commission classified sustainable development as 'development that satisfies the current needs without jeopardising the future generation's ability to fulfil theirs. It aims at incorporating the essential principles of intra-generational and inter-generational equity and has persuaded many governments to endorse the notion of sustainable development' (IPCC, 2019). The definition addresses the main concerns around the utilisation of non-renewable resources, ensuring economic growth for an increasing global population without undue impact on the environment, urbanisation and the inequality of wealth, power and opportunity (Fletcher, 2008). The meaning of sustainability is contested as there are over 300 different definitions (Dobson, 1998). In general, sustainability can be categorised into four types of capital stock:

- Human – the population, welfare, health, workforce, educational and skill base.
- Physical – productive capital such as machinery, equipment and buildings.
- Environmental – manufactured and natural resources, and biodiversity.
- Sociocultural – well-being, social cohesion, empowerment, equity and cultural heritage.

(Fletcher, 2008: 217)

Tourism is a constantly growing multi-perspective phenomenon, in which the logical consequence of growth is that the role of tourism in relation to the human contribution to climate change increases (Figure 2.1). There is a broad scientific consensus that tourism contributes significantly to climate change and different mitigation strategies have an important role in addressing emissions from the tourism sector. Carbon dioxide (CO_2), which is the most important greenhouse gas, is emitted through a range of tourism activities and, in 2005, approximately 87% of all emissions were caused by transport, 9% by accommodation and 4% by other activities (Peeters, 2007).

The first United Nations Conference on the Human Environment (UNEP) was held in Stockholm in 1972, producing an action plan for the environment based on the global environment assessment programme; environmental management activities; and international measures to support the national and international actions of assessment and management (Fletcher, 2008). Sustainable criteria defined the essential elements against which sustainability is assessed, each criterion relating to a key element of sustainability and described by one or more indicators that include ecological, economic and social aspects. A few decades later, Agenda 21 (Rio Summit in 1992) set a series of principles for guidelines

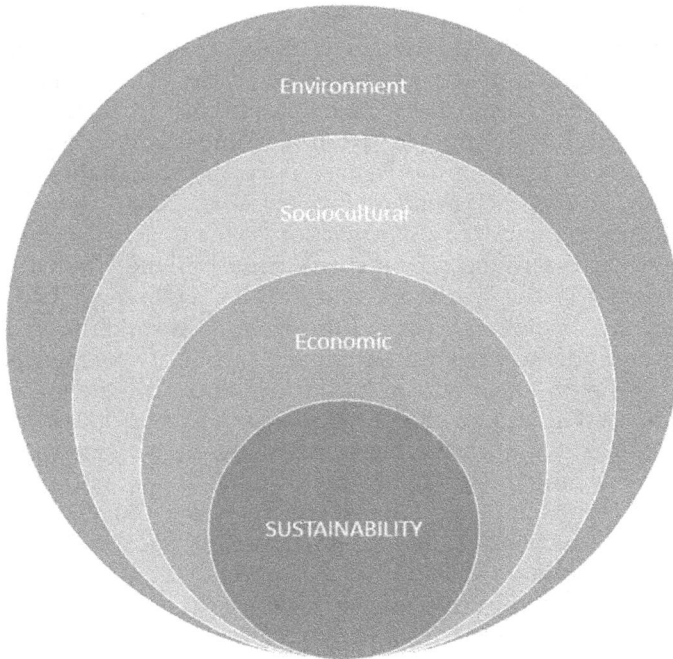

Figure 2.1 Sustainability of tourism

for environmental legislation, and global companies in raw manufacturing, design and tourism now have to meet many environmentally stringent standards. Sustainable production contributes to environmental quality through the use of natural resources, the minimisation of waste and the optimisation of products and services.

Sustainable tourism is defined as 'tourism that takes full account of its current and future economic, social and environmental impacts, addressing the needs of visitors, the industry, the environment and host communities' (UNWTO, 2005: 11). Sustainability in tourism involves economic, environmental and sociocultural aspects, which, due to the nature of tourism, could be complex and even work against sustainability. For example, economic unsustainability could develop where local investments are purely focused on developing tourist areas, resulting in structural unemployment in other areas; environmental unsustainability when the number of tourists exceeds the carrying capacity of environmentally fragile areas; and sociocultural unsustainability when the local customs and traditions undergo commercial intervention by the tourism industry (Fletcher, 2008).

The carrying capacity is defined as 'the maximum number of people who can use a site without an unacceptable alteration in the physical environment and without an unacceptable decline in the quality of experience gained by visitors' (Mathieson & Wall, 1982). When attempting to identify the carrying capacity, ascertaining the absolute number of possible tourist arrivals needs to take into account factors such as the average length of stay, the characteristics of the tourists and hosts, the degree of seasonality, the types of tourism activities, the accessibility of specific sites and the level of infrastructure (Cooper *et al.*, 2008).

Traditionally, since their introduction in the mid-1980s, sustainable tourism products have been related to ecotourism and alternative tourism, especially in the marketing and academic literature. Ecotourism in particular is linked with natural tourist attractions rather than their manufactured counterparts and sustainability is a core component of the product's definition. Ecotourism itself suggests constraints on locations in order to prevent uncontrolled development (Fletcher, 2008). The focus of sustainability in tourism has been on an actor–network perspective, on making an inventory of sustainable tourism initiatives, on the impact of transport on tourism, on sustainable tourism and transport policies and on prevailing questions in the field of sustainable tourism development (Verbeek & Mommaas, 2008).

In order to create successful interdependence between the different parties, two elements need to be taken into consideration: interdependence between humans to protect set values and the change of paradigm in its relation to nature (Helne & Silvasti, 2012). Natural capital is a complex idea that has three dimensions: critical, constant and tradable. Critical natural capital is that which is vital to life, such as the atmosphere and the ozone layer; constant capital refers to important, but not essential, elements such as a forest for a nature park; and tradable capital is that which is not highly valued and can be replaced (Barr, 2008; Davies, 2013). Ecological sustainability aims for the continuity of the planet's natural resources and concentrates on ensuring that pollution and the use of resources do not exceed certain levels. Failure to meet these goals means that future generations are left with polluted lands and wasted natural resources.

Sustainable development can also be defined in weak and strong dimensions (Davies, 2013; Närhi, 2004). The weak sustainable paradigm states that human-made capital is more important than natural capital, essentially putting human needs first (Khan, 1995; Neymayer, 2003). Strong sustainable development involves lifestyle choice approaches and an emphasis on the lack of natural resources and preserving them as long as possible (Dobson, 1998; Dryzek, 1997). These two dimensions lead to many ethical questions and have an impact on the decisions made at political and economic levels. Sustainable development is tied up in a timeframe that is essentially linked with power, thus sustainable criteria are often defined by people with the necessary financial resources (Mowforth & Munt, 2015). The question, then, is who else can define what sustainability is and how it is to be achieved?

Nowadays, the discourse of sustainable development is less debated than some decades ago when authors such as Dovers and Handmer (1992) and Lele (1991) were still critical of the concept. More recently, there has been a wide range of discussions and research on the different levels and impacts of sustainable development, for example in the writings of Gössling *et al.* (2005) and Meadows and Randers (2005). However, the debate around the sustainable development of the future is still ongoing as, according to Baptista (2014), the central element in domesticating the future through sustainability involves in particular the cultural confrontation between 'the West and the rest' and the ideology seems to be self-contradictory. On one side it promotes cultural diversity, but on the other side it operates only under a singular and homogeneous construct of the future.

It should be noted that not everyone shares the same definition of 'development' and this raises the question as to whether the 'space jumps of the wealthy' enhance sustainability in any way. It is also worth questioning if there is actually a need for such a conceptual definition of eco-social aspects of sustainable development in the same way as, for example, in environmental-social politics (Helne & Silvasti, 2012). This is because there are many approaches to defining their meanings. Sustainability has a global responsibility to simultaneously take different elements into account and focus attention on long-term local and global impacts (Salonen, 2010). Critics argue that sustainable development contains inbuilt assumptions about the need for the continued expansion of the world economy and that it fails to stress the radical changes in lifestyles and society required to overcome the problems inherent in the Western model of development. The key question, however, appears to be how to find the balance between nature and humankind.

Tools for sustainable actions

The United Nations (UN, 2015) has created an agenda for transforming our world with sustainable development by the year 2030. There are 17 different sustainable development goals and targets: no poverty; zero hunger; good health and well-being; quality education; gender equality; clean water and sanitation; affordable and clean energy; decent work and economic growth; industry innovation and infrastructure; reduced inequalities; sustainable cities and communities; responsible consumption and production; climate action; life below water; life on land; peace; justice and strong institutions and partnership for the goals (UN, 2019b).

The agenda itself concentrates on the planet Earth and states, 'we envisage a world in which every country enjoys sustained, inclusive and sustainable economic growth and decent work for all. A world in which consumption and production patterns and use of all natural resources – from air to land, from rivers, lakes and aquifers to oceans and seas – are sustainable' (UN, 2019b: 7). There is, as yet, no mention of the protection of space; this appears to remain a 'Wild West' grey area in recognised institutional global sustainable actions. However, these goals should still be treated as a moral guideline for governments and private companies to follow and act upon in activities taking place beyond Earth, even if this is yet to be stated explicitly.

The tools of sustainability include the protection of areas, industry regulation, environmental impact assessment, carrying capacity calculations, consultation techniques, codes of conduct and sustainability indicators (Mowforth & Munt, 2015). The indicators are the most recent, created at the Rio Summit in 1992, and involve elements such as resource use, waste and pollution. The indicators show the links between economic, social and environmental issues and the power relationships behind them. The formulation of sustainability indicators is a challenging task because subjectivity is introduced at each step, from the selection of indicators to their interpretation (Wong, 2006). Hence, only a few countries and regions produce systematic and standardised indicators that are capable of measuring positive and negative effects on the environment, society, the economy and culture. Others view sustainability as an objective or political consensus resulting from discussion among the stakeholders (Rametsteiner *et al.*, 2011).

Indicator-based sustainable tourism strategies are complicated by the actual process of selecting, measuring, monitoring and evaluating a set of relevant variables (Weaver, 2005). Many airlines, tour operators and travel companies have started to introduce sustainability actions and reporting schemes that act under the guidance of a certain sustainable certificate. There are also different tools and concepts for climate impact mitigation. These include, for example, improving the precision of navigation to allow environmentally preferred routes to be flown more accurately, increasing capacity at optimum flight altitudes and flexible flight planning to avoid contrail (Williams *et al.*, 2007).

Global environmental movement

Environmentalism could be described as a broad philosophy and social movement regarding concerns for environmental protection, incorporating the impact of change on the environment of humans, animals, plants and non-living matters. This 'green ideology' advocates the preservation and improvement of the natural environment and critical processes such as climate. It may be referred to as a movement to control pollution and protect plant and animal diversity, sometimes even with radical ecological anarchism (Environmentalism, 2019).

BOX 2.1 SOME OF THE KEY MOMENTS IN THE HISTORY OF THE ENVIRONMENTAL MOVEMENT

- 1962 – Publication of *Silent Spring* by Rachel Carson.
- 1970 – US Environmental Protection Agency formed.
- 1970s – Beginning of large-scale, organised environmental movement.
- 1980s – Environment generally seen by industry as a cost and compliance issue.
- 1990 – First Intergovernmental Panel on Climate Change (IPCC) report on global warming.
- 1995 – Shell forced to change disposal method of redundant oil-loading buoy (Brent Spar).
- 1997 – Kyoto protocol signed.
- 2000s – Sustainable development increasingly recognised as a business opportunity and increasing recognition that business is the key to resolving sustainability issues.
- 2005 – Major global corporations start embracing sustainability as central to strategy.
- 2015 – Paris Agreement on climate change.
- 2017 – CO_2 emissions standards for airplanes enforced by national aviation authorities.
- 2018 – IPCC report on climate change.
- 2019 – Greta Thunberg – the rise of the young generation for climate protection.

(Source: Fawkes, 2007; IEA, 2019; UN, 2019a)

Towards future sustainability

Sustainable tourism rests largely upon questioning the capitalist ideology of the Dominant Western Environmental Paradigm (DWEP) and recognising that tourism is a political and socially constructed phenomenon, in which 'some voices and agendas are heard – and others are not' (Wilson, 2015: 203). Weaver (2005) claims that sustainable development and tourism are in need of ongoing critique and deconstruction as they are some of the most contested ideas in the current academic literature.

Sustainable tourism has been slow to develop because of few short-term benefits, denial and lack of partnership building (Lane, 2009). A question could be raised about the interaction between globalisation and relationships of power (Mowforth & Munt, 2015). The key themes include the new forms of sustainable tourism and how sustainability relates to unequal development. The emergence of new forms of tourism, such as eco-luxury and those labelled as sustainable, is testimony to the identification of a problem and an attempt to signal that these options aim to overcome issues in a way that the mainstream tourism industry does not.

According to Gössling *et al.* (2009: 4), 'the most relevant issue for environmentally sustainable tourism is climate change, both because tourism is affected by climate change and because the sector is a considerable force of climate change'. Tourism is itself an important contributor to global emissions because of transportation, accommodation and other activities. Transport is responsible for 75% of all CO_2 emissions created by the tourism industry and the aviation sector alone accounts for 40% of all CO_2 transportation emissions (Gössling *et al.*, 2009). Thus, it can be concluded that tourism may never become fully sustainable as long as it involves physical movement by a vehicle that utilises a natural resource.

As the amount of air traffic increases, it can be expected that all emissions will increase with the onset of commercial space tourism. Ross *et al.* (2010) claimed that 1000 new suborbital launches, using a hybrid rocket engine, will cause significant changes in the global atmospheric circulation and distribution of ozone and temperature. Their study, published in *Geophysical Research Letters* and funded by NASA and the Aerospace Corporation, simulated the impact of 1000 suborbital launches of hybrid rockets from a single location, calculating that this would release a total of 600 tonnes of black carbon into the stratosphere.

The results were that the resultant layer of soot particles remained relatively localised, with only 20% of the carbon straying into the southern hemisphere, thus creating a strong hemispherical asymmetry. They predicted that this unbalance would cause the temperature to decrease by about 0.4°C in the tropics and subtropics, whereas the temperature at the poles would increase by between 0.2°C and 1°C. The ozone layer would also be impacted, with the tropics losing up to 1.7% of its ozone cover and the polar regions gaining 5–6%. The findings were not a precise forecast of the climatic response to a specific launch rate of a specific rocket type, but demonstrated the sensitivity of the Earth's atmosphere and the possible impact of mass commercial space tourism (Ross *et al.*, 2010).

Because of the unique nature of their combustion chemistry, rocket engines emit large amounts of black carbon when compared to, for example, a modern jet engine. More recent research by Ross and Sheaffer (2014) demonstrated quite unexpectedly that such rocket launch emissions could actually contribute to cooling Earth's lower atmosphere and surface despite previous research indicating the opposite, bringing a new aspect to the impact of rocket exhaust on the climate. However, further scientific study is needed to fully understand the long-term implications for the climate.

Climate Change

The idea of climate change is not new as it has long been known that carbon dioxide released into the atmosphere can harm the Earth. The earliest calculations date back to the 1800s when Swedish Nobel prizewinner Svante Arrhenius used the greenhouse effect to explain his theory that fluctuations in glacial warm periods depend on fluctuations in atmospheric carbon dioxide. At that time, his calculations did not raise alarm as he speculated that it would take 3000 years before human activity would raise the level of carbon dioxide by 50% and even then, it would only result in future generations being able to enjoy a more pleasant climate and better harvests.

In the late 1950s, American Charles Keeling started to systematically monitor the amount of carbon dioxide in the atmosphere and found that it actually increased faster than any previous estimations, resulting in the Earth warming up faster, with potentially devastating consequences. However, these concerns remained for a long time only within the scientific community, as did the internal research carried out by oil companies, showing similar results. The impact of climate change became an international political issue much later. However, in spite of a series of international politically-oriented meetings, the levels of emissions have risen year by year (Seuri, 2018).

Intergovernmental Panel on Climate Change

The IPCC (2018) report drew together everything scientifically known about global warming at 1.5°C for the first time. All countries in the world must considerably tighten their emissions targets if we want to limit global warming to 1.5°C, as even a 2°C warming would severely exacerbate the consequences. Here are some examples:

- More people would suffer from lack of water.
- Hundreds of millions of people would be more exposed to poverty, especially in developing countries.
- More than 10 million people would be put at risk due to rising sea levels.

The aim of the report was to find out the practical meaning of the 1.5°C objective of the Paris Climate Agreement and included a summary for global decision makers. The key point of the summary was that if the emissions created by humans can be greatly reduced in the next few years, it is still possible to limit global warming to 1.5°C. Achieving this goal requires fast, far-reaching and record-breaking emissions cuts (IPCC, 2018).

> ## BOX 2.2 THE ROLE OF THE INTERGOVERNMENTAL PANEL ON CLIMATE CHANGE
>
> - The IPCC does not do the research itself but gathers and evaluates scientific information produced by research institutes.
> - The international climate policy, such as emission restrictions, relies on IPCC climate reports.
> - A 1.5°C report was commissioned by the Paris Climate Conference in 2015.
> - The report has over 90 authors and has been commented on 42,000 times.
> - The report refers to a total of 6000 studies.
>
> (Source: IPCC, 2019)

Urgent action is needed to keep warming since the pre-industrial era to 1.5°C, as even an increase of more than half a degree would mean moving from dangerous to very dangerous climate change with effects accumulating in a way that would be extremely difficult to predict. Resetting net emissions means that humankind would only produce an equal amount of carbon dioxide emissions as the seas, soils and growing forests. CO_2 emissions, especially among wealthy industrialised nations, should turn sharply from 2020 if the 1.5°C target is to be reached and global net emissions should be reduced to zero by about 2050 (IPCC, 2018).

Even though a half a degree difference between climate goals may sound trivial, the impact could be dramatic. For example, a 1.5°C temperature increase in the Arctic Ocean would mean an approximately 10 centimetre rise in the sea level in this area with a particularly sensitive sea ecosystem. This would have a big impact on the infrastructure of coastal areas, on coral reefs important for maintaining fish stocks, and it would also increase rainfall and hotspots. Reductions in emissions should be fast, far-reaching and record-breaking and, to accomplish this, our lifestyles, especially in Western societies, should become environmentally friendly, without dependency on fossil fuels, in all aspects over the next 10 years, particularly in the use of goods and recycling. Even though the timescale may seem short, the same period of time has seen the introduction of social media and the launch of the first iPhone, both of which have brought significant change in the way people communicate nowadays (Koistinen, 2018; Seuri, 2018).

There are different pathways to dealing with reducing emissions, most of them relying on the assumption that carbon dioxide can be captured, for example, from a plant barrel with technology and stored on the Earth's crust (IPCC, 2018). However, these technologies have so far remained at the pilot stage. Maintaining the 1.5°C target would mean, for instance, that the share of renewable electricity, such as solar and wind power, would rise to over 80% of global electricity production and that there would also be a significant increase in nuclear power. The situation can be facilitated by the rapid development of technology and the recent decrease in the cost of producing renewable energy (Koistinen, 2018).

The lifestyle of a typical consumer is likely to change markedly as the need to use oil reduces drastically, leading to the electrification of traffic and greater reliance on the public transportation network. However, emissions reductions in transport are among the most difficult, as the world is still heavily dependent on fossil fuels, even in Western societies such as the United States. Consumer behaviour should be directed towards choosing products with a lower carbon footprint and avoiding those that create higher emissions. One example, the production of cattle and dairy products, is responsible for high levels of methane emissions. Consumers should cut emissions by reducing the overheating of and heat loss from homes and avoiding flying – all symbols of the Western lifestyle and possibly a bit painful to give up. However, achieving the 1.5°C goal would significantly reduce future human suffering and protect nature (Alston, 2019).

Emissions

The European Commission (2018) suggested that the European Union (EU) should cut emissions to zero by 2050 to reach the 1.5°C target of the Paris Agreement to stop global warming. It means big changes in the member countries, such as in the use of electricity. Currently, 25% of the electricity within the Union is still produced with coal, but in the future, 80% needs to be produced in a renewable way and 15% by nuclear power, resulting in the need for more funding of €175–€290 billion per year. Emissions trading within the EU will be a cost-effective way to reduce emissions, as well as cutting emissions within countries as cost-effectively as possible (Peljo, 2018).

China's emissions in particular have increased dramatically during this millennium, explained by the country's fierce economic growth: gross domestic product (GDP) grew by around 10% every year in the 2000s. In China, economic growth and carbon-based energy production have gone hand in hand, tripling its CO_2 emissions. However, particularly because of visible pollution in the big cities, China has started to make emissions reductions and, for example, has invested in electronic traffic. India's emissions have also increased considerably and how this is tackled in the future will be an essential consideration in reducing the global carbon dioxide emissions. There have been investments in solar power, the most competitive form of energy on both sides of the equator (Heikkinen, 2018).

It needs to be remembered that even though China currently produces the most carbon dioxide emissions, the United States has contributed the most to global emissions over time. Its total emissions have doubled. China, Russia, Germany, Great Britain and Japan come after the United States. It is remarkable that all of them have used more of the global carbon budget than India. Oil-producing countries are also at the forefront as its production creates emissions. If there is no methane recovery in the oil fields, then emissions when burning oil are high. That is why emissions are high in Russia and the oil-producing countries of the Middle East, while in Norway, for example, methane is largely recovered. In many oil-producing countries, oil is also used relentlessly (Peljo, 2018).

BOX 2.3 THE HISTORY OF EMISSIONS CREATED BY HUMANS

There have been four stages in the history of the emissions created by humans:

- The initial phase was before the Industrial Revolution, when the main sources of energy were wind and hydropower.
- The second phase started with the Industrial Revolution of the 18th and 19th centuries. With the steam engine, humankind learned to harness its fossil fuels and increased carbon dioxide emissions. However, long-term emissions growth was very slow. A clear leap forward took place in the late 19th century, when the fast-growing United States began to industrialise, and oil was harnessed as an energy source.
- The third clear phase began after the Second World War, when, as a result of reconstruction, emissions accelerated, especially in the Western world and the Soviet Union.
- The fourth breakthrough came at the turn of the 21st century, after which emissions from emerging economies, especially China and India, have increased dramatically.

(Source: Heikkinen, 2018)

Population Growth

The human population has increased from 1 billion in 1804 to approximately 7.7 billion people in 2019. This growth is a driver for much of the sustainability debate as it has major consequences on the environment regarding natural resources and energy use. Most of the world's inhabitants (4.4 billion) currently live in Asia, dominated by China with 1.43 billion people, followed by Africa (1.2 billion) and Europe (0.7 billion). Latin America and the Caribbean Islands are next (0.6 billion), then North America (0.4 billion) and Australia and Oceania (0.04 billion). It has been estimated that between 2017 and 2050, the population will double in 27 African countries, whereas it will face a decline in Western societies such as those in Europe. The forecast for the population in 2050 is 9.8 billion and 11.2 billion people by 2100 (Hiltunen, 2019; UN, 2019a).

A population explosion has an enormous impact as it puts a strain on the environment. It threatens the future of sustainable life on Earth, with the largest environmental effect being global warming, followed by the greater demand on natural resources, the destruction of natural habitats and the increased production of waste (Mittal, 2013).

The Rise of the Middle Class

Despite the megatrend of globalisation, which has equalised operations for economic growth (through global tools such as the internet and crypto currencies), there remains an existing imbalance between the wealth of developed and less-developed countries, and billions are still denied a life of dignity (UN, 2019b). However, the percentage of individuals living on less than $2 per day has dropped from 37% in 1990 to less than 10% in 2015, and there has been growth of the middle class, with 3.8 billion people now belonging either to the middle class or even wealthier economic groups (Hiltunen, 2019; Roser & Oritz-Ospina, 2017).

The growth of the middle class is forecast to be particularly prominent in formerly less-developed countries such as China and India, with a projected population of almost 5 billion people by 2030. The rise of the middle classes will have a major and sustained impact on the global economy by accelerating economic growth and methods of consumption, as well as contributing to increased usage of the limited natural resources of the Earth. The middle classes also participate in governmental decision-making by being active voters and guiding mass-powered decision-making in their consumer choices (Hiltunen, 2019).

Energy Consumption

Energy consumption on Earth in 2017 was five times greater than in 1950, with 35% of the energy produced from oil, 28% from coal and 24% from natural gas (Hiltunen, 2019; Ritchie & Roser, 2018). The acceleration in energy consumption has been driven by the global growth in electricity and gas demand, but has been especially spurred by sustained economic growth and rising demand in China, which has been the world's largest energy consumer since 2009. China's energy consumption has been driven by power generation, strong individual demand and increased transportation fuel consumption encouraged by a growing vehicle fleet. In 2018, the total energy consumption in the United States reached a record high of 2.4 Gtoe, up by 3.5% on 2017, partly driven by its drastic summer and winter weather conditions. On the contrary, the EU managed to decrease its energy consumption by –1%, and in Germany alone by −3.5%, due to decreasing consumption in the power sector, a milder winter climate and energy efficiency improvements (GESY, 2019).

Energy consumption in the transportation sector is forecast to increase at an annual rate of 1.4% from 104 quadrillion British thermal units in 2012 to 155 quadrillion units in 2040 (EIA, 2019). Worldwide, petroleum and other liquid fuels are the dominant source of energy for transportation, although their share is forecast to decline from 96% in 2012 to 88% in 2040. In aviation, energy efficiency improvements have been more rapid than in most other sectors of passenger transport. However, while energy efficiency improved by 3.2% per year between 2000 and 2014, it slowed to less than 1% per year between 2014 and 2016. In addition, CO_2 emissions have continued to rise and accounted for around 2.5% of global energy-related CO_2 emissions in 2018 (IEA, 2019).

Since 2000, while the demand for air transport has more than doubled, the aviation sector has achieved significant energy efficiency improvements. Better energy efficiency has been pursued as a means of improving profitability, fleet renewal and aircraft utilisation. The rise of low-cost airlines increased the number of passengers per flight, lowering energy use per passenger (IEA, 2019). From 2021, the increase in emissions due to international flights will have to be compensated through carbon offsetting. There is a strong need for purposeful integration of the disciplines of transportation research and engineering to develop more renewable energy technologies and fossil fuel-free efficient transport systems (Becken, 2015).

Oil

Oil is involved in everything in one way or another in modern societies and is considered to be the most important and strategic non-renewable resource that cannot be substituted with anything else at its current volume (Hiltunen, 2019). The idea of a peak in oil production was first introduced in Hubbert's (1956) peak theory, estimating that in any given geographical area, from an individual oil-producing region to the planet as a whole, the rate of petroleum production tends to follow a bell-shaped curve (Hubbert, 1962). Since then, many scientists have attempted to predict a global peak in this finite resource and to predict the future of oil supply and demand (Becken, 2015).

In 2018, approximately 100 million barrels of oil were used per day. Worldwide consumption has more than doubled in the last 50 years and the demand has increased significantly in rapidly industrialised Asian countries such as China and India (Ritchie & Roser, 2018). It is important to remember that oil is not just a raw material, but a central factor in the global economy and political decision-making. Former oil crises have caused resource wars and recession due to increased inflation and have contributed greatly to global greenhouse gas emissions.

There are no official estimates of the total oil consumption in tourism, but global carbon dioxide emissions can be used as a proxy for oil, with transport being responsible for 75% of the carbon footprint of tourist activities. Globally, 8000 billion kilometres were travelled by tourists in 2005, equalling 10 million trips to the Moon and back. The largest share was by air (49%) (Becken, 2015).

With contemporary technology, virgin stocks of several metals also appear to be inadequate to sustain quality of life in the modern developed world and so there is a need for increased efficiency of material usage, substitution of materials and improved recycling (Fawkes, 2007). New energy solutions are also needed alongside non-renewable energy sources for the sake of the environment. Sun, wind, hydropower and biofuels may provide a practical solution for the future (Hiltunen, 2019).

Increased Climate Change Awareness

Postmodernism helps to capture the high degree of difference and fragmentation lying at the heart of contemporary cultural change (Harvey, 1989). It is closely linked to the emergence and growth of the new middle classes following the same global trends and products. It offers a position refusing to acknowledge the way in which commodities are used and understood by different social groups and communities. This has resulted in a highly polarised and simplified debate concerning the most appropriate way of holidaying.

Following the IPCC (2018) report and increased public awareness of the impact of climate change, the younger generation in particular, inspired by the environmentalist and the school strike promoter Greta Thunberg, has become globally sensitive to environmental concerns, creating a new megatrend of making daily life decisions based on their carbon footprint. The younger generation's attitude has since spread to other generations and people have started to become more aware of responsible ways to travel in order to minimise one's carbon footprint. General social norms have also started to determine what kind of travelling is socially acceptable and even desirable, and at the most radical extreme, some have even chosen to no longer travel for environmental reasons.

There have also been changes to how businesses operate, as they deal with consumer demands for more sustainable actions, and are forced to increasingly focus on the efficient use of resources. Businesses are becoming involved in more circular economy-based solutions. In Jensen's (1999) *The Dream Society*, he writes about the future of business. His theory has different dimensions, one of them being the market for adventures including extreme adventures and overcoming one's fears. The conscience of the economy can be supported through charity projects, and taking responsibility for business actions is becoming a major factor in competitiveness. Following an environmentally sustainable path, businesses voluntarily take part in social and environmental respon-sibilities as a part of their business development plan and stakeholder relationship. Some of the most comprehensive approaches to sustainable marketing follow core principles such as ecological orientation and a long-term, life cycle approach, as well as consideration for the wide-ranging impact of production. Recently, businesses have become very visible in promoting sustainability as a part of their business philosophy and large global corporations targeting the younger-generation buyers, such as IKEA and H&M, have moved towards strategic proactivity.

BOX 2.4 LEVELS OF COMPANY SUSTAINABILITY

- **Rejection and non-responsiveness:** The company is simply ignorant of sustainability issues and is actively opposed to sustainability concepts.
- **Compliance:** Includes the vast majority of companies, meaning that they try to actively reduce operating risks and follow corporate responsibility standards in order to protect the company's reputation.
- **Efficiency:** Already involved in actions which include introducing sustainable policies to reduce costs, but these are often operated outside the core business activity. The company needs, therefore, to make sure not to miss any strategic opportunities to create sustainable values if the link to operational action and the company is not clearly seen by the consumer.
- **Strategic proactivity:** The corporation seeks competitive advantage by engaging with different stakeholders in product development.
- **Sustaining:** Includes companies that offer pioneering alternative interpretations of business values and success.

(Source: Fawkes, 2007)

In the future business environment of commercial space tourism, sus-tainability will most likely be seen as a 'hygiene factor', meaning that it is a necessity to incorporate it into the business plan. 'Early adopter' space tourism enthusiasts will likely have a high awareness of current megatrends, including sustainability issues. Many are likely to come from influential and business backgrounds, and will be sensitive to consumer criticism on the individual actions taken by the board of management (Fawkes, 2007).

Case Study: Remote Area Considerations

With its remote and harsh living environment, the Arctic region of the Earth has similarities to 'destination space' and locations for potential future colonies. It can be used as an example to mimic a space setting, and similar issues to those raised in some of the recent discussions around environmental and security concerns in the Arctic could, in theory, replicate in future in the space environment. While tourists and private travel companies tout the benefits of visiting the Arctic, especially in terms of increasing one's environmental awareness, several bodies of research indicate that those benefits have been quite minor or even non-existent, whereas visiting such remote regions has environmental consequences. The benefits largely accrue to the tourists while the region bears the environmental costs. In this light, a critical question to be addressed is whether space tourism will be any different to the current Arctic exploration that satisfies the tourist's desire for self-exploration.

The dialogue around indigenous people and fairness to the local community cannot obviously be translated directly, as space has no known 'indigenous inhabitants' to protect from 'human colonialism'. However, some future space tourism launch sites may be increasingly based in the Arctic regions, such as Kiiruna (Sweden), in order to provide access to polar orbital tourism. Already, the increased volume of visitors experiencing terrestrial space tourism, such as the global tourism megatrend of wanting to witness the Northern Lights, has had a negative local impact in terms of more congestion, environmental damage and noise. It has also conflicted with the traditional source of livelihood for indigenous people, such as the Sami in Lapland. Some reindeer pastures are now traversed by adventure-seeking tourists.

There were significant changes during the 2010s in the north Arctic region, as the Arctic become the 'latest trend'. Many nations, even those not in the Arctic region, such as China, started to pay considerable attention to the new possibilities the area offers and fuelled the paradigm shift of international security. Global climate change is one of the reasons that the Arctic has started to raise interest, as it provides more economic possibilities by shaping the risk perceptions of all states, including those of the great powers and militaries (Kopra, 2017). For example, the Northwest Passage and the Northern Route will in the future provide more open waters for cargo ships to carry goods from Asia to Europe. There are also new emerging tourism sectors, such as the cruise industry, that have taken an interest in operating in the Arctic region not yet accessed by the tourism industry.

China's increasing interest in the Arctic has given rise among Arctic countries to some security questions about its motives (Kopra, 2017). According to Hastings (2014), Chinese investments in Icelandic industries raise some concerns about the vulnerability of being dominated by a country that is in size, power and influence much greater than Iceland. In Greenland, the government has also been keen to attract Chinese investment to develop its extractive industries in areas such as mineral resources (Mouyal et al., 2016).

When evaluating what security means in the Arctic, military bases, such as those located in Greenland, have traditionally been some of the most visible markers. Their presence has raised questions about the environmental damage caused by military 'protection' and practices. The environmental impacts of the military in peace time traditionally rise from the use of fuel, the land, water areas and air space, the use of different toxins, the thermal impacts, nuclear and other accidents, as well as being the cause of CO_2 emissions (Heininen, 2017). Pollution generated by military activity has had far-reaching consequences on human societies' inability to address the challenge of global environment change and degradation. As a result, there has been an evaluation and elimination of accumulated ecological damage in the Russian Arctic Zone as there are unsatisfactory levels of atmospheric air in a number of places, caused by current extractive industries and the military.

In their article 'Toxic Splash: Russian rocket stages dropped in Arctic waters raise health, environmental and legal concerns', Byers and Byers (2017) claim that Russia has used the Arctic water as a dumping place for their 'toxic splash' caused by falling space and surveillance rockets. The rockets, designed to land back in Arctic waters, have raised various concerns for locals. Twelve launches took place between 2002 and 2017 at the 'Rockot' satellite launcher with SS 10 intercontinental ballistic missiles. Even though a 'Notice to Airmen' was given before the launches, they caused aerosol clouds, environmental debris impacting the area's large colonies of sea birds, polar bears and marine mammals, and feelings of insecurity in the local inhabitants (Byers & Byers, 2017).

In order to mitigate these environmental impacts, various international committees have established environmental strategies, such as those addressed at the annual Arctic Circle Assembly, to conduct strategic action assessments to protect the Arctic environment. The strategies include planned clean-up operations, assisted also by the military, the prevention of further environmental pollution, the improvement in the quality of the environment and the conservation of biological diversity. In one example, a geopedological survey expedition of contaminated islands of the Franz Josef Land Archipelago performed experimental work to eliminate representative sources of negative impact (Shevchuk, 2017).

From the tourism point of view, the increasing global temperatures caused by climate change have created new opportunities for some private operators, such as the cruise industry, to start operating in Arctic waters. This phenomenon has resulted in new environmental and safety discussions between national security bodies, dealing with marine tourism security, and commercial cruise liners. Commercial cruise ships are likely to carry several thousand passengers, many of them elderly, making it challenging to safely evacuate in the event of a sudden emergency. According to Pincus (2017), the US coast guard has already expressed concern that it may not be possible to rescue all passengers in a safe way as the Northwest Passage has many isolated areas with no infrastructure. In some regions, roads and basic infrastructure are available, but in a scenario with hundreds of sudden casualties and only a small local hospital with, for instance, five available beds, this will not secure passenger safety. These sudden emergencies also place the local inhabitants in danger if their only health-care facility becomes burdened.

The US coast guard has started to teach some private sector companies, including those from the tourism industry, how to act in case of emergency, using the latest technological aids and solutions such as drones, which, in emergency situations, could be sent to 'eye-witness' the level of help needed and investigate the area in general (Pincus, 2017). The US coast guard's actions are in line with other Arctic security trends, such as the Arctic and High North Project seeking to broaden and widen the understanding of security and insecurity in the region. The partners range from international lawyers, environmentalists and strategists to policymakers and military practitioners and reports have suggested an increase in high-tech Arctic drone bases (Rogers, 2017). Along with ice-breakers and drone bases, it may be essential to form a virtual net over the Northern Sea Route. Activity in the Arctic region will increase, highlighting the importance of search and rescue capabilities, servicing and security. All new infrastructure needs to be able to secure and control this new 'Suez' passage of Europe–Asia trade (Kennedy-Pipe, 2017).

To conclude, there are noticeable similarities between some of the environmental and security issues that have recently occurred in the North Arctic and the current weak signals in space exploration, and in the formation of a new adventurous sector of tourism:

- Firstly, China is making vast investments both in the Arctic and space (the Moon), indicating early signs of area 'colonisation'. Such actions have already created concerns related to local security and formed a discussion around fairness and equality of ownership among locals.
- Secondly, private industry operations have been increasingly challenged – obtaining rights for private tourism industry actors to operate in the Arctic area is an example. New forms of tourism may cause, in the case of an accident, vast damage to already fragile environments and raise ethical questions around health and safety for tourists, as medical assistance may not be possible on-site.
- Thirdly, Arctic areas have suffered from environmental damage caused by military actions. For example, the Russian Arctic waters have been used as satellite landing sites for some decades now, causing disturbance to both local people and the environment. As the creation of the US Space Force already implies, the space environment will increase military actions in the future with different impacts, possibly even leading to resource or area protection wars in space.

- Fourthly, the latest technological innovations and solutions will have a major role in enhancing both the quality of life in the region and short-term visits. For example, new satellite networks will increase the level of health and safety security in the Arctic, as well as in space.

Conclusion

Since the discovery of fossil fuels, there has been an unprecedented rise in the well-being of societies. However, the world we live in can be very cruel not only to the environment, but also to our social values and the systems based on them. Following the release of the IPCC (2018) report on climate change, we are now facing quite radical social reforms to correct the course of environmental destruction. Population growth is causing increased energy and oil consumption, leading to even more emissions. The opening up of commercial space tourism in the current climate of environmental crisis certainly works against the global sustainability megatrends.

It is important to note that as climate change requires long-term work, the consideration of short-term financial and other goals will be an important factor in negotiations between businesses and governmental bodies. Even though the negative impact of climate change may still appear far away, the cost of action is paid now. The negative effects of climate change are also most pronounced in poorer countries, which have less weight in global decision-making than their wealthy counterparts.

Governments are in a key position to tighten emissions trading screws and use existing tools such as carbon taxes or certain business support to promote sustainable development. A key aspect is that if societies understand the depth of change, it becomes more possible to implement it within the framework of a democratic society. As the Arctic area has harsh environmental conditions that in some parts could mimic the space environment, the region could be used to enhance future sustainable planning as an Earth-bound study example to replicate levels of environmental impacts and other security concerns faced in the remoteness.

Sustainability in the tourism industry could be achieved by engaging the industry, governance and regulation, researching market benefits and the role of the media industry in opinion forming and using the ideas developed in social sciences to explore decision-making and social trends. The focus should be on sustainable science and this is to be achieved by understanding the sociology and psychology of the new types of tourism behaviour.

3 Futures Forecasting

I think fundamentally the future is vastly more exciting and interesting if we are a spacefaring civilization and a multi-planet species than if we are not. You want to be inspired by things. You want to wake up in the morning and think the future is going to be great. And that is what being a spacefaring civilization is all about.

Elon Musk, 2017

Introduction

The process of creating and visualising an image of leaving planet Earth may seem both mysterious and challenging due to many combined factors, such as changes in society and the timescale of technological innovations. When planning new futuristic sectors of the tourism industry, our ideas could remain as imaginary concepts, or the current travel trends could potentially be enhanced and commonly adapted over a sustained timescale. As commercial space tourism is only just taking its first steps, this chapter explores the background of futures research, to examine how to approach a travel activity still commonly thought of as a dream.

The chapter first defines the characteristics of futures research, and then explains some of the elements of environmental scanning and the life cycle of weak signals. New future forecasting models, a Futures Map (Kuusi *et al.*, 2015) and the Sustainable Future Planning Framework (Toivonen, 2017), are then introduced and explained with examples relating to space tourism, along with a future scenario case study on the establishment of space colonies.

Reading Future Signals

Homo sapiens evolved in Africa during a time of dramatic climate change 300,000 years ago. Like any other early humans, their behaviours evolved through reading signals from the environment that helped them to respond to the challenges of survival in unstable conditions. They began to settle down in villages, which later turned into success-driven synergised cities, to ensure the continuity of the species into the future (Smithsonian, 2019).

Fast-forwarding to the present, the IPCC (2018) report raised questions about the future impact of climate change on habitable areas and, in some radical future visions to ensure human survival, it could be advantageous to begin developing space colonies (Hawking, 2016). Over the past decades, scientists and academic scholars have increasingly noted the benefits of human expansion into space. Collins and Autino (2010) were among the first to connect the involvement and development of commercial space tourism with reducing the danger of human extinction on Earth as a result of disasters such as nuclear wars, climate change and asteroid impacts. The prospect would depend in large part on the availability of low-cost space travel which appears to be achievable only through the development of a vigorous space tourism industry (Collins, 1994).

In order to anticipate such future changes in developments, it is essential to look for emerging issues and weak signals from society and to practise environmental scanning, which is a process of acquiring information. Weak signals are emerging ideas, intentions, discoveries and innovations that are not yet trends, but have the potential to impact upon local areas within three to five years (Smyre & Richardson, 2015). Futurist thinking helps one to understand how changes occur and how they are predicted. In the beginning, when a change emerges, few people have an awareness of it. Eventually, the new trend affects the everyday lives of most people and everyone becomes familiar with it. At its foundation, anticipating a change involves stable factors, weak signals, trends, megatrends and wild cards (Hiltunen, 2013).

Bell (1997: 73) characterised futures research as such: 'the purposes of the future studies are to discover or invent, examine or evaluate, and propose possible, probable and preferable futures'. There are two main approaches to the research, the first one being a scenario approach and related methods for the examination and evaluation of alternative futures. In this approach, the future scenario may be described as a specified path connecting the present state to at least one picture of the future and creating answers to questions such as how a hypothetical situation might develop step by step and what alternatives exist to each step for facilitating or preventing the process (Kahn & Wiener, 1967).

The other approach uses Delphi surveys that evaluate the probability and preferability of some specified futures, using a professional panel with expertise related to the future issues. These two paths may occasionally be combined to use some versions of Delphi surveys and form a scenario funnel describing a range of possible futures looking forward from today's standpoint (Kosow & Gassner, 2008; Kuusi et al., 2015). The elements of a future strategy process may also integrate the 'phases of foresight', including strategic intelligence, sense-making, selecting priorities and implementation, to promote the internal validity of the futures process (Cuhls *et al.*, 2014).

International tourism research has traditionally reflected on previous tourism research in order to visualise the future. The forecasting of new travel phenomena has tended to be based on the history, current trends and perspectives of researchers (Ryan, 2018). Space tourism represents a new era of innovative start-ups, electric cars and the 'show-off' experience of sharing via social media platforms. Consumer values in society have turned into megatrends, with a lessening emphasis on material possessions and an increasing concern for experiential and quality of life issues (Yeoman, 2008). The world of experience represents the new luxury, as consumers seek to experience something that is hard to access but not necessarily owned. New trends can be identified in entrepreneurship, such as collaborative consumption-based virtual platforms such as Airbnb and Uber. Current megatrends in tourism promote sustainability and its transformation into trans-modern eco-luxury.

Environmental scanning

Weak signals and environmental scanning have been discussed in the futures literature since the late 1960s by Aguilar (1967), followed by the different aspects of environmental futures by Ansoff (1975). A distinction between weak signals and emerging future issues can be made with the concept of the future sign, in which weak signals are understood more as signals of emerging issues (Hiltunen, 2007). Choo (2006) explains that environmental scanning analyses information about every sector of the external environment that can assist in planning for future paths.

BOX 3.1 THE LIFE CYCLE OF WEAK SIGNALS

The life cycle of weak signals and scanning the environment include different stages:

(1) **Idea creation**: Artistic works, doctoral dissertations, alternative press, patent applications, specialised journals, social media mentions.
(2) **Elite awareness**: Insider newsletters, research reports by analysts, scientific journals, business leader magazines, popular intellectual magazines, shared social media experiences.
(3) **Popular awareness**: TV programmes, radio programmes, newspapers, opinion surveys, fiction and non-fiction works, social media sites.
(4) **Government awareness**: Government financing and sponsoring of reports and surveys, public discussion forums, draft legislation.

(5) **Procedural routinisation:** Government policies and regulations, education curriculums, instruction manuals to staff, industry cooperation.

(6) **Record keeping:** Legislative records, historical records, analysis studies, government archives.

(Source: Adapted from Choo [2007] and Hiltunen [2007])

Future road mapping can be used as a tool to provide an extended look at the future, composed of the collective knowledge and imagination of visionaries to predict developments and decision-making (Matos & Afsarmanesh, 2004). Road mapping may be described as a structural, yet flexible, tool when navigating within a pool of uncertainties. Different future phases on the road map can act as guidance for decision makers in order to be able to forecast future needs and requests. Traditionally, scenarios have been developed to support the formulation of the most desired vision, but at the same time there has been criticism that this is too distant to support strategic development. Linking scenarios with road mapping, however, can initiate an exploratory and creative phase to understand and catch uncertainties. Scenarios open up to more than one future which can be equally plausible, whereas road mapping can provide a framework for condensing all information into one map and timeframe – revealing windows of opportunities and thus linking decision-making with future scenarios (Rikard & Borch, 2011).

Future Forecasting Models

Strategic thinking about the future should be incorporated into tourism planning as a means to continually process external and internal information and adjust to changing situations (Wilkinson, 1997). Sustainable futures planning has become a requirement in the tourism industry and importance is given to research designed to model a sustainable tourism future (Mojic & Susic, 2014). Within the tourism industry there are already some sustainable development future strategies, such as sustainable indicator-based strategies, which protect the environment that is impacted on by tourism with a range of measurements and practical actions. However, to enhance the futures forecasting and planning process, more tourism-oriented future research tools need to be created to assist with the process. Accordingly, two new future forecasting models, the Futures Map and the Sustainable Future Planning Framework, are introduced below.

Futures Map

The Futures Map is the comprehensive description of the outcomes futures research proves. It comprises all relevant pictures of the future identified during the process and all relations between these pictures and between them and the present state. It also provides assessments about time frames, desirability and possibility of these pictures.

(Kuusi *et al.*, 2015: 62)

The Futures Map includes all possible futures characterised by two parameters or dimensions as identified during the research process (Figure 3.1). The first dimension is when the future is assumed to realise (time of realisation) and how users will appreciate its realisation (desirability and preferred futures). The other dimension is the possibility, meaning whether the future may, according to the best current knowledge, realise or whether it is only an abstract option, such as a utopia, not fitting reality.

During the futures mapping process, the actor (or organisation) obtains information that changes its expectations of the ranges of desirability and probability of possible future developments. It is based on envisaging, assuming that the more desired the vision is, the more likely that future path is to occur. The acceptable futures are those that are desirable but not bad futures that the people of the shared vision believe they can avoid. The Futures Map can also include different quality criteria to improve its pragmatic validity, especially useful for companies planning for the desired future path of their enterprise. There are six validation criteria: Criteria 1 and 2 identify the most relevant or important possible futures; Criteria 3 and 4 cover casually relevant facts by their identified futures; and Criteria 5 and 6 concentrate on the needs of key customers (Kuusi *et al.*, 2015).

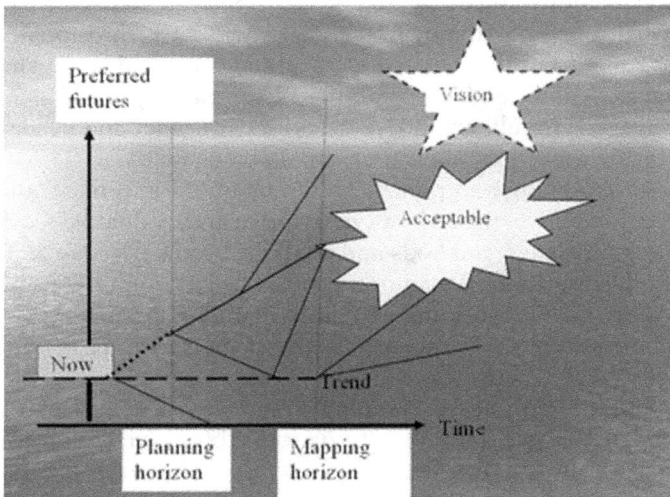

Figure 3.1   Futures Map

Most importantly, especially when forecasting space tourism, the Futures Map is based on two concepts: the planning horizon, focusing on the history and current situation of space tourism, and the mapping horizon focusing on the future visions, with both time horizons being defined during the framing process. The planning horizon can be connected to the concept of a road map, as during the time of framing, the involved actors are committed to follow the specified road of the map. A road map requires a vision, a horizon time frame and a horizon goal, and at the end of the planning horizon new evaluations of the vision and the acceptable futures are needed. The mapping horizon is the anticipated horizon of the map of the possible futures and a scenario path may in fact have ended, already reaching the mapping horizon. Most scenario paths are defined by the mapping horizon and there can be many scenario paths leading to the same end point of the mapping horizon (Kuusi, 2017).

Sustainable Future Planning Framework

The Sustainable Future Planning Framework was created by the author in 2017 to act as a future forecasting tool and presented in her article for the *Finnish Journal of Tourism Research*, called 'Sustainable Planning for Space Tourism' (Toivonen, 2017). The intention was also that this new future model could be easily modified with the empirical findings discovered later on in the research process in order to be able to add more in-depth knowledge and perspectives on the discourse of sustainable space tourism.

The Sustainable Future Planning Framework (Figure 3.2) can assist in foreseeing the likely implications of human actions and their impacts on future sustainability and in creating feasible possible futures. In a similar way to other tourism planning models, it provides forward-looking planning and identifies the principal elements of sustainable development. The framework indicates that 'planning, sustainability, weak signals and future scenarios' should act in synergy with each other and hence formulate the future aspects of space tourism (Toivonen, 2017).

Tourism planning and policy can be organised into five traditions – normative, predictive, procedural, descriptive and evaluative – offering a mixture of knowledge, methods and analytical tools (Dredge & Jenkins, 2011: 14). The procedural tradition provides advice on how to plan and manage tourism. The Sustainable Future Planning Framework is placed under this tradition, as it targets specific elements to act in synergy to achieve sustainable planning.

The Sustainable Future Planning Framework concentrates on the future sustainable planning of a futuristic subject, such as in this case future space tourism. In the framework, the 'planning' involves governmental legislation and action plans, scientific understanding, as well as modelling and databases. Duval and Hall's (2015: 451) planning

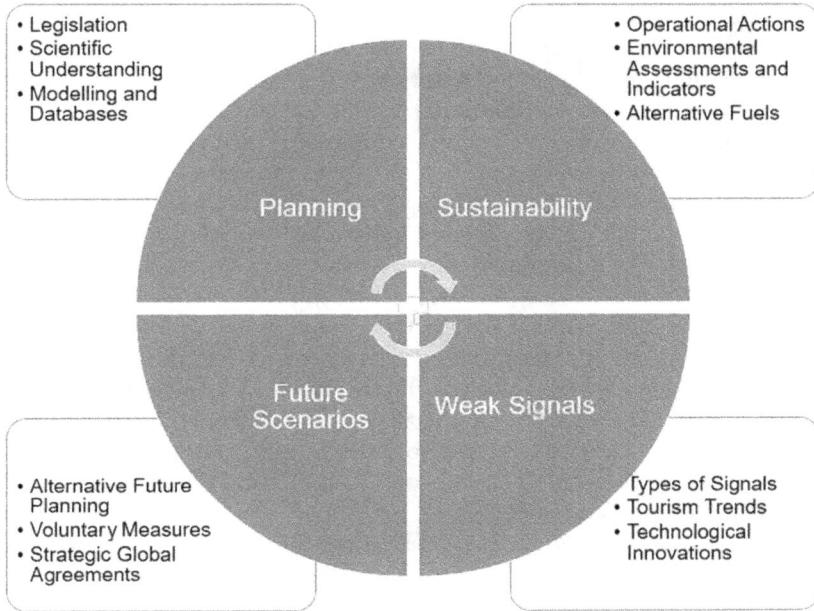

- Legislation
- Scientific Understanding
- Modelling and Databases

- Operational Actions
- Environmental Assessments and Indicators
- Alternative Fuels

Planning

Sustainability

Future Scenarios

Weak Signals

- Alternative Future Planning
- Voluntary Measures
- Strategic Global Agreements

- Types of Signals
- Tourism Trends
- Technological Innovations

Figure 3.2 Sustainable Future Planning Framework

discourse for future space tourism points out that a critical question in the policy implementation process should consider: 'Who will be in charge of enacting global policy and governance, the future agreements developed by the political and private sectors, and is there actually a mutual willingness to leave some parts of space as tourism locations completely untouched?'.

Tourism planning and policymaking are also a result of the ideas, actions and collaboration of diverse agencies and draw from many disciplines including politics, economics and history (Dredge & Jenkins, 2011). In this context, space tourism could be viewed as an objective or political consensus resulting from discussion among stakeholders (Rametsteiner *et al.*, 2011). Butler's (1980) Tourism Area Life Cycle model of destination development elements could assist in policy planning for space tourism as it clarifies the different stages of destination development, thereby acting as a guideline for the development process. Saarinen's (2014) considerations on the impacts of larger global economic situations and political relations might also benefit policy formulation; as the identity of a destination, such as in this case the space environment, is a changing product of transformation, the postmodern values of tourists and the current sustainable goals of the tourism industry have an impact on the policy planning processes.

'Sustainability' in the Sustainable Future Planning Framework involves operational actions, environmental assessments and indicators as well as alternative fuels. Indicators of sustainability are different from traditional indicators of economic, social and environmental progress measuring changes in one part of a community as if they were entirely independent of the other parts. Sustainability indicators reflect the reality that the three different segments are very tightly interconnected. For example, air quality has an effect on human health and stockholder profits. Indicator-based sustainable tourism strategies are complicated due to the actual process of selecting, measuring, monitoring and evaluating a set of relevant indicators (Weaver, 2005).

Existing sustainable indicator measurement base examples are available within the tourism industry. For example, the aviation industry practises sustainable actions by focusing on a more energy-efficient fleet and voluntary carbon offsets paid both by the industry and the customer (Broderick, 2009). Sustainable indicator-based international agreements, such as the Kyoto Agreement (2013) and the UN Paris Agreement (2015) on climate change, have been targeted at strong governmental actions to reduce increased global emissions.

There is ongoing research on alternative fuels in the space tourism industry, such as rubber-based propellants, that may lower the impact of emissions (Carter et al., 2015). There is, however, a strong need for purposeful integration with the disciplines of transportation research and engineering to develop more renewable energy technologies and fossil fuel-free efficient transport systems (Becken, 2015). Future spaceship models could, for example, operate by using renewable fuels or solar power.

'Weak signals' in the Sustainable Future Planning Framework include types of signals, tourism trends and technological innovations and have certain characteristics such as surprise, visionary discoveries and acceptance of the trend through learning and developing (Coffman, 1997). In generative dialogue about the future, as well as adaptive planning, continuous transformation is not focused on what already exists, but on what is emerging. Uskali's (2005) theory of weak signals distinguished four types of weak signals: 'feeling, uncertain, almost certain and exact' signals. The space tourism industry has suffered from technical development problems, particularly those related to ensuring passenger safety, in the implementation process, but issues have been researched as they have appeared. As the infrastructure for space tourism has now been created, the industry has passed the 'feeling' and 'uncertain' stages and is heading towards the 'almost certain' stage, as most of the infrastructure needed is now available for passenger usage. Weak signals thus suggest that sustainable research and indicator forming to assist the policymaking process urgently need to take place to ensure a sustainable beginning for this new tourism sector.

The space tourism industry has been developed to reflect current adventure tourism trends. Even though the idea of the creation of space tourism works against the current trends of sustainability, the implementation of the industry could incline even more towards trans-modern eco-luxury. Actions are already practised by other tourism sectors, such as aviation, to make even a small step in the process more sustainable and environmentally friendly. Space tourism fits within the wave of current trends in technology seen in electric cars and innovative start-ups. The space race decades ago had already enabled the public to benefit from different technological innovations developed for both astronauts and inhabitable space stations. Future space vessels may enable more comprehensive research to inform the design of experiments on longer-term physiological changes due to space flight. These discoveries could eventually help the human species to colonise the Moon or Mars (Caplan et al., 2017).

In the Sustainable Future Planning Framework, future scenarios concern alternative future planning, voluntary measures and strategic global agreements. Future scenario planning is a strategic planning method that can be used to make flexible long-term plans (Kahn, 1965). Pauwels and Berger (1964) defined contractual futures of impossible possibilities to be part of the planning scenarios as there could be four generic alternative futures leading to continued growth, collapse, social discipline or transformative scenarios. The future scenarios could, for example, involve sustaining values to avoid a collapse and planning for radical options, such as the development of space colonies.

In the space tourism sector, some voluntary measures have already been taken, for example, to ensure that operational-level safety and security, as well as global agreements, such as the UN Paris Agreement (2015) on climate change, are followed. However, there needs to be a further global legal framework to decide on the most prominent issues, such as the definition of outer space and the demarcation of a boundary between outer space and airspace, which is currently governed by different legal regimes, but not commonly agreed upon. Space tourism will be a private sector participator in the space environment besides the previously exclusive national and military usage, meaning many legal issues will need to be resolved quickly or alternatively voluntarily pursued. These issues include, for example, voluntary guidelines for space debris, equitable access and use of the geostationary orbit commonly used by communication satellites, a mechanism for the settlement of disputes and a regulated flow of technologies (Sharma, 2011).

Future Scenario: Space Colonies

Let's now explore a future scenario in which space colonies are built for human settlement and use a Futures Map (Kuusi et al., 2015) as a tool to indicate the time horizons for practical actions to occur at present, or to determine whether the idea remains a radical vision and utopia. The case study is divided into two sections: the planning horizon exploring the history and the present, and the mapping horizon, introducing more of the imagery of a human space settlement.

Planning horizon

In the *Journal of Acta Astronautinca*, Collins and Autino (2010) presented a theoretical analogy to understand the current human predicament, named 'Pregnant Earth', and to explore the potential of space travel to revitalise human civilisation. The analogy suggests that humans' terrestrial civilisation is 'pregnant' and dangerously overdue with extraterrestrial offspring. The theory explains that once a human civilisation in space is safely born, the current stresses on the mother civilisation will be cured and the new life may eventually even surpass its parent. It states that the development of an extraterrestrial civilisation will lead to a wide range of activities outside the planet's ecosystem and the evolution will solve not just material problems, by making the vast resources of near-Earth space accessible, but also help to cure the emptiness of modern commercial culture causing a decline in public morality and society's goals and challenges.

This needs critical consideration, though, as the notion of the drive to explore is also often invoked especially by pro-space advocates; can the journeys of millionaires, such as Dennis Tito, to the International Space Station be justified as the drive to settle space? It may be that they not only fail to cause humankind to band together, as often mentioned by Collins (2003, 2014), but could in fact distract from solving pressing terrestrial issues. It could be asked, for example, why one should worry about income inequality if asteroid mining would make all humans rich, and why one should continue to worry about climate change if space-based solar power is going to meet all of our energy needs many times over (Spector, 2020).

An opening of a new geographical frontier with all its challenges could, however, offer enormous value for the cultural growth of modern civilisation and create optimistic visions of an unlimited future. It is even pointed out that 'implementing the "Pregnant Earth" agenda can prevent the cultural regression and start a true world-wide Renaissance' (Collins & Autino, 2010). The desire to explore the solar system can be traced back to the 15th century, when Europeans began to explore new frontiers to search for new trading routes and allies to expand their economic and political power. The current era of space exploration has so far concentrated on gaining a better understanding of the planet Earth and our place in the universe. Robotic exploration has offered the main possibility to thoroughly investigate the soil of Mars and the dark side of the Moon. The use of artificial robotics has many scientific advantages without a limited mission duration, the high cost of life support and the return transportation involved in manned missions. This has been particularly important as the exploration of the solar system has expanded to encompass new research fields such as exobiology in order to understand the conditions leading to the creation of planets and the emergence of life (ESA, 2019a).

The world of science fiction has, however, presented visions of space colonisation since 1869 when *The Brick Moon* was published, describing an inhabited artificial satellite (Hale, 1869). In the 20th century, the idea of communities in space can first be seen in *Beyond Planet Earth*, by Konstantin Tsiolkovsky, followed by many authors around the world to date. The need for space colonisation to reverse-colonise Earth and act as a backup for human civilisation to improve the survival prospects of the human species in the event of a global nuclear war was first officially recognised by US governmental officials and related professionals, such as physicists, during the Cold War (Morgan, 2006).

To date, no space colonies for permanent human habitation off planet Earth, except the International Space Station, which has had a permanent human presence for several decades, have been built, as the building of a space settlement has presented enormous technological and economic challenges. Different organisations are involved in space colonisation, such as the Space Studies Institute and the National Space Society (USA), funding the study of space habitats and maintaining scientific articles and books on space settlements. The dream of establishing a long-term human settlement on Mars has a supporting project with SpaceX developing extensive space flight transportation infrastructure, and the Artemis Project aims to set up a private lunar surface station. The British Interplanetary Society promotes ideas for the exploration and utilisation of space including Mars, and for locating other habitable areas.

In the first phase of establishing a space colony, the technology needed to control ecological life support systems must ensure space settlements meet the essential needs of colonists. Secondly, it is necessary to seriously consider the psychology of humans: how would people behave in such a place in the long term (Chaikin, 2015)? The stage of establishment also involves a range of other topics for consideration, such as space traffic management, the building of infrastructure and property rights and ecological matters (Collins & Autino, 2010). For example, at the very beginning, it should be decided globally if it is ethically and legally right for the material to manufacture the settlement to be taken from the Moon or Mars (Johnston, 2017). Also, when visiting other planets, all robotic spacecraft should be sterilised to prevent the risk of contaminating not only life-detection experiments, but also the planet itself. Human explorers are also vulnerable to bringing contamination back to Earth if carrying other planetary micro-organisms.

In 2006, the first experimental space station module, Genesis I, was built by the private commercial space flight company Bigelow Aerospace (2019). Another attempt to build a self-sufficient colony was Biosphere 2 in Arizona, built between 1987 and 1991, which attempted to duplicate the Earth's biosphere and serve as a centre for research, teaching and life-long learning about the Earth's living systems (Biosphere 2, 2019). Space agencies, such as NASA and ESA, have built test beds for advanced life support systems, testing the duration of the effects of lengthy human space flight, though not necessarily on permanent living. The construction of the Mars Research Station began in 2001 with the mission to help develop key knowledge needed to prepare for human exploration of Mars. It is a laboratory for learning how to live and work on another planet and has a prototype habitat that will land humans on Mars and serve as their main base for months on the planet. Professionals such as engineers, geologists and human factor researchers have been able to live for months at a time in isolation in a Mars analogy environment on locations on Earth that provide an inhospitable climate comparative to that on Mars (MDRS, 2019).

One of the major environmental concerns of the current time is the increasing consumption of the Earth's resources to sustain our way of life. As more and more nations make the change from agricultural to industrial societies, the standard of life will improve, resulting in more people competing for the same resources. The fact that humanity is currently limited to one planet and its resources creates a problem in spite of inventions and technology to allow more prudent usage of the Earth's resources. Space colonies could therefore provide an answer to this dilemma for future generations. Such colonies could become a reality once there is a solution to the medical problems caused by microgravity (weightlessness) and the high levels of radiation to which astronauts are exposed after leaving the protection of the Earth's atmosphere (NASA, 2019h).

The limits to the supply of Earth's natural resources, such as oil, has already created resource wars between countries and some experts are of the opinion that expansion into near-Earth space will be the only alternative to end or prevent such wars and to preserve human civilisation. Opening access to the unlimited resources in space could therefore facilitate world peace and security. The continuity of the economic profits to be made from natural resources is also becoming more uncertain and one of the ways to address the situation is to expand into space. The same rockets currently used for fighting for the resources on Earth could in theory be used to travel to space and bring back all the needed reserves. There could be high returns from extraterrestrial settlements if there are investments in low-cost orbital access (Collins & Autino, 2010).

However, critical consideration needs to be given to the fact that no one has yet coherently explained how these resources will possibly benefit the whole of humankind rather than a handful of rich companies and countries. Rather, the creation of the Space Force in the United States indicates that outer space is not likely to solve terrestrial warfare and inequality, but instead greatly exacerbate those dynamics (Spector, 2020).

The solar system alone has enough natural resources and energy to support anywhere from several thousand to over a billion times that of the current population on Earth, providing opportunities for both resource mining and the building of a workforce colony, leading to new economic opportunities (Lewis, 1997). However, to access these planets, there needs to be new revolutionary methods of travel, such as interstellar travel faster than the speed of light. Luckily, precious metals and even water resources can also be found on asteroids and planets that are close to Earth. In the future, these could act as fuel stations for space travel to destinations further away.

The future offers the new private space businesses an increasing opportunity to extend their activities in space. However, if their only motivation is the attempt to achieve monopolistic control and profit, they may actually start hindering any development in space. The Space Renaissance Initiative is a programme to accelerate the expansion of human activities in space by advocating investment to reduce the cost of space travel and it has been joined by a number of space-related organisations. Reducing the cost of space travel is especially important in the generally poorer state of the world's economy in order to create more employment in the space sector, increasing its contribution to the world's economy (Collins & Autino, 2010).

Mapping horizon

Over the years there have been various proposals and designs for future space settlements, many styled with the ultimate science fiction imagination. In 1977, NASA published a design study guide, *Space Settlements: A Design Study for Space Settlements*, to provide a city planning policy guide on what future colonies should look like. It focused on orbital civilian habitats, town-sized space stations housing tens of thousands of civilians each, which follow the real-world laws of physics. It was suggested that the design of the space colonies could be urban, walkable, transit oriented, dense and inclusive, exactly what urbanists advocate for Earth, and which would be physically possible to build if space becomes widely accessible or more heavily settled. The design of the orbital space station town would be circular, and it would spin to simulate gravity (O'Neill, 1976).

The guidebook covers everything from the likely sociology of the people inside to how much air a colony would need and the kind of rockets that could service it. There is even a comprehensive town plan defining how much space would be needed for residents, schools, transportation and other infrastructure. The population estimation would be

15,000/km² and there would be one major mass transport system such as a monorail or moving sidewalk to connect different residential areas in the same colony. Parks should be high enough (50 metres) to prevent the feeling of claustrophobia, and to allow a true feeling of being in the open as there would be no escape from the surrounding walls (O'Neill, 1976).

Another infrastructure design of a space colony, suggested by the British Interplanetary Society think tank (Hollingham, 2013), would consist, similarly to NASA's 1970s outline, of a vast hollow cylinder which would rotate to provide gravity for the people living on the inside. There are already practical designs made for a ring-shaped colony where the cylinders are 6.5 kilometres in diameter and 32 kilometres long, capable of holding up to 10 million people. Anthropologists such as John H. Moore have estimated that the population of the first colony could be around 150–180 people, which would permit the existence of a stable society for over 60 generations, over 2000 years. Initially, the population would survive with as few as two women if human embryos were available from Earth, and use of the Earth's sperm bank would hinder inbreeding.

The provision of various materials is necessary to build up a space colony, including access to water, food, radiation protection and simulated gravity. Other essentials would include construction materials, energy production, transportation vehicles, communication methods, life support and recycling of raw materials (Hickman, 1999). Solar energy would also be available in space without any blockages to sunlight, providing a reliable source of energy, and thermal power plants would be needed to meet the electrical power needs of space settlers. There could even be wireless power transmissions of special solar cells or power beams converted with high efficiency to send power to different locations on Earth with zero emissions, thus also eliminating future greenhouse gases and nuclear waste. However, it is important to note that, in accordance with the second law of thermodynamics, all energy must be released as heat and the energy beamed to Earth will eventually be released in the form of heat, and therefore space-based energy cannot really be attained without also exacerbating warming on Earth (Spector, 2020).

In future, colonies on the Moon, Mars or asteroids could extract local materials, such as helium, argon, iron and titanium, as launching materials from Earth would be costly (Perlman, 2009). Cryptocurrencies and blockchains could be used as a viable method of exchanging payment for the resources between future colonies and the Earth. The human colonists could mine the Moon and some minor planets and build beamed power satellites that would supplement or even replace power plants on Earth. It could be possible to take advantage of the variety of raw materials, unlimited solar power and microgravity. Colonists could use other methods to create products that may not be possible with Earth's atmosphere and gravity. In addition to potentially replacing the existing Earth-polluting industries, colonies could also help the Earth's environment in other ways. Since colonists would inhabit completely isolated manufactured environments, they could refine our knowledge of the Earth's ecology (NASA, 2019h).

One of the main limitations to the establishment of space colonies and space resource exploitation has so far been that they are very high-risk investments, as they have never been done before. There is an enormous cost for initial investments as all the infrastructure needs to be built from scratch and the expected return of investments may take decades (Lee, 2003). As the timescale for building new space colonies can be considered a long one, anything from the most optimistic vision of 20 years (Elon Musk) up to 100 years (Stephen Hawking), there needs to be long-term planning on global government initiatives and with private corporations providing sufficient funding.

The key players in the establishment of colonies would be private companies, such as SpaceX, Blue Origin and Virgin Galactic, developing spacecraft that provide cheap and reliable access to space. If there is rapid progress in that sector, the first colonies could be built in 10 years' time and be finished 20 years from now. The first colonies could be floating in space rather than attached to the Moon, already supported by the achievements of the current technology (Stone, 2017).

Economics plays one of the most important roles in making the creation of space colonies possible, with the ultimate requirement being that the methods of travel and launch systems become much cheaper than currently available. SpaceX Falcon 9 rockets have so far been one of the most cost-effective space vehicles presented in the space industry, with a launch costing about $56 million. As SpaceX is currently developing a reusable launch system development programme, this could encourage more space-based enterprise, leading to even further cost reductions. If SpaceX (2019) continues to be successful in developing reusable technology, this may have a major impact on the cost of access to space and be a game changer for the timescales of establishing a space colony.

Location-wise, the Earth's Moon would provide a practical starting point for the first planetary colony as it is already familiar and located relatively close to the Earth, enabling a smoother transfer of humans, goods and services between the two places. The Moon also has some surface gravity, which would provide some comfort outside a built infrastructure, as well as a temperature suitable for the human body, although there is no atmosphere to provide protection from space radiation. Various nations, such as the United States, China and India, have already stated that their future space exploitation plans will involve a settlement on the Moon, especially as a means to assist longer distance space travel, with the Moon acting as a 'pit stop'.

Private industry space operator SpaceX has established future colony plans on Mars as 'The Moon is too small without atmosphere, Venus is too high pressured with acid hot baths, Mercury too close to the sun and

Jupiter and Saturn located too far from the Earth' (Musk, 2017). At a distance of 229 million kilometres from the sun and about 150–300 days from Earth (depending on various factors), Mars as a planet offers some gravity (37% of the Earth's), an atmosphere primarily composed of carbon dioxide with some nitrogen and argon, a manageable temperature for the human body (–140°C to 30°C), a similar day length to Earth (24 hours 37 minutes) and a land mass 97% of Earth's (Musk, 2017; NASA, 2019i).

SpaceX's Elon Musk has published a paper on his vision for the colonisation of Mars, unveiled in the journal *New Space* (2016) and titled 'Making Humanity a Multi-Planetary Species'. At the centre of the vision is a reusable rocket and spaceship combination, called the Interplanetary Transport System (ITS), with its booster being the most powerful rocket in history. The ITS boosters will launch many spaceships and fuel tankers into orbit over the course of their potential lives, and the rockets are designed to fly about 1000 times each. The spaceships will stay in orbit and then depart en-masse when Earth and Mars align favourably once every 26 months. This type of architecture could make the colonisation of Mars more affordable and get up to one million people to Mars within the next 50–100 years (Musk, 2017; Wall, 2017).

Outside the solar system there are millions of potential stars that could be possible targets for future colonisation. To reach such destinations, interstellar travel with travel times at the speed of light and a technologically advanced spacecraft would be required. In science fiction-styled visions, there could even be generations-long voyages on 'colony ships' with independent energy generation. The ancestors of the original crew would arrive at the final destination to expand the human race beyond the Milky Way (Schmidt & Zubrin, 1996).

BOX 3.2 PERSPECTIVES THAT SUPPORT SPACE COLONISATION

- The first colony could be counted as the Earth's first act of self-replication, enabling space manufacturing and allowing a further increase in colonies, while eliminating costs and dependence on Earth.
- It would satisfy the human drive to explore and discover.
- There is no known life in space, meaning no indigenous habitat suffering from human colonisation.
- Overpopulation of the Earth needs new solutions to replace non-renewable resources.
- If Earth's resources are replaced with those from space, in theory the Earth's population may no longer need growth limitations.

- The Earth's environment can be spared by moving some industry and also human inhabitants to space colonies.
- It would ensure the survival of *Homo sapiens* if the Earth became uninhabitable in the case of a sudden catastrophe such as an asteroid impact.

(Source: McKnight, 2003)

However, critical examination is necessary: can the processes that have caused issues in the first place, such as overpopulation and over-extraction of Earth's resources, simply be transferred and extended to a new region, as has already previously been done in many places on Earth such as the Arctic? The flaws in the system are likely to remain and possibly amplify. History has shown that conquests of remote areas in order to source new resources and areas for settlements have often caused inequality and even wars – would it be possible for humankind to prove itself different this time in space?

Conclusion

Throughout history, it has been essential to practise environmental scanning and interpret weak signals emerging from society in order to ensure the prosperity and even the survival of humans. Current global megatrends already relate to the discourse and desired actions of environmental protection, with weak signals suggesting the increase in unconditional values and actions to prevent any further destruction of the environment among the younger generation. Radical futures scenarios even envisage that it is a necessity to begin developing space colonies to ensure the survival of humans.

As sustainable planning has become a major requirement in today's tourism industry, there is a strong need to create new future models to assist in designing and forecasting more feasible sustainable tourism industry futures, as these have further reflections on society. Future models, such as the Futures Map (Kuusi *et al.*, 2015) and the Sustainable Future Planning Framework (Toivonen, 2017) introduced in this chapter, could be further supplemented and utilised as tools for designing operational strategies for actions towards a sustainable space tourism industry. For example, the future formation of space colonies was explained in this chapter with the assistance of the time horizon predictions placed on the Futures Map.

There is a strong signal that there is an urgent need to assure future generations by all possible means that sustainable actions are to be taken very seriously and included in all cooperative strategies between different governments and private companies, to maintain future credibility among the younger generation. Gaining this understanding will be crucial, especially for commercial space tourism companies as they are to start operating in an era of global environmental climate crisis.

4 Planning Sustainable Space Tourism

> Future space tourism presents an important philosophical challenge that can be harnessed for sustainability, forcing participants to consider their place in the universe, relationship to other beings, and especially concepts of time.
>
> Carter *et al.*, 2015: 457

Introduction

One of the current megatrends in the tourism industry is sustainability. Sensible sustainable planning should always be undertaken prior to any development of tourism activities and projects (IPCC, 2018; Mojic & Susic, 2014). Given that tourism was originally framed as co-existing with its environment, research concentrating on new resources and survival planning is capable of facilitating new, environmentally based discussions in space tourism discourse. In developed Western economies in particular, the approaches to tourism planning and policy have been linked to profound changes in ideological and sociopolitical landscapes (Dredge & Jenkins, 2011). Governments have tended to offset their responsibilities through commercial agreements with external providers, such as corporates with financial power, which the future space tourism industry with its affluent owners will certainly represent.

This chapter explains the background to sustainable tourism planning, using destination life cycle models as an example. The different links between the space tourism industry and sustainable development are then explored by providing insight into operational, cultural, economic, resource and survival levels of action in the future of sustainable space tourism.

Tourism Planning

Tourism planning is about predicting and requires some estimation about the future. Gunn (1979) was among the first to define tourism planning as a tool for destination area development and Getz (1986) classified it as predominantly a project based on solving a problem. Planning and policy research has recently developed from the growing influence of critical research approaches and post-disciplinary perspectives focusing attention on explaining tourism planning processes (Dredge & Jenkins, 2011).

There are five traditions in tourism planning: boosterism; an economic industry-oriented approach; a physical/spatial approach; a community-oriented approach; and a sustainable approach (Hall, 2011). The physical/spatial approach, in which tourism acts as a division of space by allocating specific activities to specific areas, and the sustainable approach, in which the provision of long-lasting and secure livelihoods reduces resource depletion and environmental degradation, could in particular present suitable features for sustainable space tourism planning. This is because the sustainable approach offers tools for the sensible development of a new industry from the very beginning and the physical/spatial approach concentrates on the activities taking place in specific areas, such as building up a spaceport hub. The interdependence of humans through the protection of set values and the paradigm change in their relation to nature need to be taken into consideration and it is important to acknowledge the importance of natural politics and decision-making in evaluating their environmental, social and cultural impacts (Helne & Silvasti, 2012; Salonen, 2010).

Sustainability Planning

The knowledge gained from reflective practice and attention to ethics and values has stimulated critical theoretical developments in planning sustainable tourism. For example, the framework of Dredge and Jenkins (2007) for understanding tourism policy and planning, presented a conceptual analysis for treating the developing knowledge of tourism planning as an iterative process, including the institutional context, issues, drivers and influences, actors and agencies, and policy dialogues as essential elements in understanding future policymaking (Dredge & Jenkins, 2011).

Tourism usually interacts and overlaps with a range of other policy areas such as transport, regional development and environmental management. The policies and actions aimed at implementing sustainable space tourism should therefore be situated within a broader policy framework of which tourism is only one component. Regulations for the tourism industry are often a combination of many sources and an effective partnership and rapport between these governmental sources will impact on the future formation of sustainable space tourism (Dredge, 2015; Mowforth & Munt, 2015).

The sustainability discourse itself assumes objectivity towards environmental issues and the intellectualisation of travel. However, sustainable tourism suffers from a theoretical fragility, not only calling into question its universal applicability, but also leading to a specific focus on resource conservation and protection in tourism. It has even been claimed that the development of sustainable tourism has failed to fulfil expectations and has even acted as a barrier to development as a result of its theoretical weakness (Sharpley, 2015).

Once the commercial space tourism industry is properly established with completed safety regulations, similarly to any other business, there is a high chance that it will start developing progressively. The beginning of the industry will be relatively small scale, including a costly pioneering phase, with wealthy elite travellers, but the scale of the activities will eventually start growing while the industry matures. With time, the space tourism industry may even have the potential to become a mass market business, similar to aviation now.

Timescale Development Comparison

When considering its historical timescale, the pioneering stage of space tourism can be compared to the era of the Vikings seeking a new land in harsh Arctic conditions without any existing knowledge of the outcome and under extreme stress. Even though the image of the 'rich elite' popping into space for 10 minutes to take a selfie from a window with the curvature of the Earth in the background may not reflect this metaphor, there will, if technological advancement continues, be future pioneer tourists on the surface of the Moon and even on Mars.

When space tourism starts moving away from an activity solely for the 'minor elite' to something desired by the 'mass elite', this next phase could be compared to the historic Grand Tour in Europe, when the wealthy wanted to explore and educate their children through personal experience of the destination. Once the advent of reusable space rockets and advanced technological solutions cuts production costs, the space tourism industry could move on to a phase similar to that of early aviation. The first commercial space flights will be very short suborbital flights, similar to the United States' and the Soviet Union's space race flights of the 1960s.

The first space adventure tourist paid over $20 million for his orbital adventure departing from Russia. However, if space tourism reaches the mass market the most radical estimations forecast the price to be similar to current long-haul first-class flight tickets, making space tourism affordable for many travellers. Weaver (2005), however, argues that the majority of new destinations may never reach the mass market, and if the cost does not rapidly decrease, this may be the case for the space tourism industry as well. In the near future, space tourism is expected to become a multibillion-dollar business targeting engineers, scientists and entrepreneurs as well as the general public.

In 2004, newly established space tourism companies were given an extended learning period, which allowed them to develop their rockets and space vehicles without federal regulations in the United States. Congress even hailed the passage of the Space Act preventing the Federal

Aviation Administration from stunting the growth of the industry and paving the way for entrepreneurs and businesses (Davenport, 2015).

Destination Life Cycle Models

An important question is to determine when 'destination space' actually starts to properly form. Will this definition also require a physical infrastructure such as a near-Earth orbit tourist area with space hotels and other associated tourist facilities, or can destination space form itself around just being in the physical location of space? Many previously remote wilderness destinations and resorts on Earth have become overcrowded, usually implying a considerable deterioration in the original attractions. While it might not be implausible to explore the implications of the rapid growth of space tourism, every individual destination is likely to face its own issues of capacity and congestion that underlie thinking about the tourist destination life cycle (Cole, 2015).

One of the paradoxes faced by the space tourism industry in its development phase will be trying find a balance between the environmental impact and customer requirements. Either way, if the development of space tourism remains only for the minor elite and thus offers 'joyrides' for the ultra-wealthy, or if it extends to a much larger market segment and then exerts a more significant environmental impact, the creation of a new space tourism product, from the environmental point of view, cannot really be justified (Spector, 2020). However, as the commercial space industry has now formed and is about to be an active player in the sector of adventure tourism, sensible and sustainable-oriented future planning should be emphasised to avoid environmental mistakes, by learning from previous knowledge of tourism destination planning.

Visitor markets and destinations overlap in the heuristic tourism literature, such as Butler's Tourism Area Life Cycle (TALC), which incorporates ideas from Plog's life cycle model about the composition of a destination into the product life cycle (Cole, 2015). The core idea of Plog's (1974) model is that the destination life cycle corresponds to the evolution of markets from allocentric to psychocentric. The allocentric discovers the destination and growth is experienced as first near allocentrics and then allocentric-leaning midcentrics visit. However, once the threshold between allocentric-leaning and psychocentric-leaning midcentrics is passed, the destination will enter a decline, as it can draw visitors from an increasingly smaller pool of tourists (Plog, 1974). However, Plog's model has been contested by the suggestion that each market drawn to a destination can actually form their own unique relationships and, as such, a destination could be seen to exist at multiple stages and many psychocentric destinations actually appear to be thriving (McKercher, 2005).

Butler's TALC is a model developed in the early 1980s explaining the stages involved in the development of a tourism destination. The TALC model identifies six stages in the life cycle of a tourism destination. After reaching the stagnation stage, there are five future scenarios between the stages of rejuvenation and decline:

(1) **Exploration**: A small number of tourists; primary natural or cultural attractions; no secondary tourism attractions; tourism has no economic or social significance to local residents.

(2) **Involvement**: Emergence of secondary tourism facilities; a tourism season may develop; pressure develops for governments to improve transport for tourists; local residents become involved in tourism.

(3) **Development**: Local involvement and control of tourism decline rapidly; high numbers of tourists may exceed the local population during peak periods; heavy advertising creates a well-defined tourism market; external organisations provide secondary tourism attractions; natural and cultural attractions are developed and marketed; local people experience physical (even negative) changes to the area.

(4) **Consolidation**: Tourism growth slows down, but exceeds the local population; the area's economy is tied to tourism; marketing is wide-reaching; major tourism chains and franchises are presented; resort areas have a well-defined recreational business district.

(5) **Stagnation**: Visitor numbers have reached their peak; carrying capacity has been reached or even exceeded; tourism causes environmental, social and economic problems; artificial tourism attractions overtake the original primary attractions; area has a well-established image, but is no longer fashionable.

(6a) **Rejuvenation**: Requires a change in tourism attractions; previously untapped tourism resources may be found.
  (a) Successful redevelopment leads to renewed growth.
  (b) Minor modifications to capacity levels lead to modest growth in tourism.
  (c) Tourism is stabilised by cutting capacity levels.
  (d) Continued overuse of resources and lack of investment lead to decline.
  (e) War or other catastrophe causes immediate collapse in tourism.

(6b) **Decline**: Unable to compete with newer tourism attractions, holidaymakers are replaced by weekenders or day-trippers; hotels may become flats for residents or retirement homes; ultimately, the area may become a tourism 'slum'.

(Source: Butler, 1980; Cooper *et al.*, 2008)

In space tourism, the number of tourists, as measured in Butler's (1980) TALC model, will not automatically reflect the success of 'space' as there are certain limitations regarding reaching and living in the destination. Access to space can currently be accomplished by a space vessel designed for transporting a limited number of people. The success of the destination in this case is more bound to the favourable political and economic climate that enables investment in technological solutions, providing access to and ensuring suitable living conditions in space, rather than to the interest in the destination itself (Toivonen, 2017).

The involvement stage is bound up with the future global legislation set for space exploration. If the trends become in favour of privatising parts of space and letting private companies have access to areas previously considered a common treasury, allowing them, for example, to build infrastructure on the surface of the Moon, then the development stage of space tourism may increase rapidly. To enable infrastructure, suitable future technological discoveries are an obvious necessity, and environmentally friendly ways of operating them should play an important part in this technological planning process. The current pioneering stage provides an opportunity to plan for, and focus on, controlled sustainable alternative tourism, which could at a later stage be expanded to the concept of sustainable mass space tourism (Toivonen, 2017).

The access, cost, infrastructure, technological solutions and safety factors all have an impact on the consolidation of space as a destination. If all the targets related to these factors are successfully met, space as a destination may even reach the stagnation stage, where the original facilities become old and run down, leading to space as a destination either to rejuvenate or face decline. Cole (2015) claims that these stages of the TALC embody the ideas of congestion and sustainability, and considering the volume of visitors anticipated, many space destinations – orbital, moon based or otherwise other-worldly – are likely to become overcrowded.

Saarinen (2014) questions the meaning of a decline or a new cycle of rejuvenation in the development of a tourist destination, and whether it is just the result of internal practices or the outcomes of development actions. The attempt to follow the regional development discourse for space tourism creates an obvious challenge as space is a vast concept and not a traditional destination with characteristic landmark features to guide the implementation processes. However, it also does not have any known local inhabitants to take into consideration, which simplifies the process (Toivonen, 2017). Learning from the examples of successfully completed rejuvenation cycles in some environmentally fragile areas, such as Arctic towns, could lead to the creation of tools that are needed for future scenario planning in the development of space tourism. Ma and Hassink (2013) similarly claim that new evolutionary perspectives on tourist destination development should focus on analysis of the changing characteristics of, and especially the underlying reasons for, destination development in a changing operational environment.

In his article 'Space Tourism: Prospects, Positioning and Planning', Cole (2015) forecasted that the future of space tourism with unlimited growth might be experimental, with mistakes made along the way, gradually moving into an infinity of accessible destinations and spaces. Most likely there will be some slowdown in the rate of increase and the same travellers will make multiple trips. Later on, the suborbital flight may be the primary mode of intercontinental travel, and businesspeople may travel regularly to corporate subsidiaries on the Moon where wealthier tourists have purchased timeshares and recreational homes. An Airbnb website may be adapted to space – Spacebnb.com. It is also pointed out that in the more distant future, as more permanent colonies and dedicated off-Earth tourism destinations evolve, one could presume that there will be similar confrontations to those portrayed in several science fiction movies. There are, therefore, many lessons to be learnt from many tourist destination experiences, past and present, that should be incorporated into thinking about future space tourism (Cole, 2015).

Linking Space Tourism and Sustainability

> Any kind of tourism can be responsible if it adheres to the principles of responsibility, requiring the responsibility of stakeholders involved in the travel activities and the achievement of the sustainability goal requires that operators take responsibility for the action and their consequences.
> García-Rosell, 2019

Sustainable development for future space tourism may at first appear to be an impossible task to accomplish due to the elements commonly associated with both tourism and space travel, as trendy practices in tourism concentrate on keeping nature and destinations as untouched as possible (Mowforth & Munt, 2015). However, in the tourism industry and within tourist activities, the status of 'becoming sustainable' already begins when the tourism business is not only concerned about its economic future success, but also wants to introduce environmental and social aspects as part of its future operational activities and business strategies. Nowadays, it is actually a major business risk not to include sustainability in the business model as it can cause market perception risk, regulatory risk and social pressure risk (Fawkes, 2007).

The concept of combining space tourism and sustainable development is still a new idea, but the two themes can be linked on five different levels:

- Operational level
- Cultural level
- Economic level
- Resource level
- Survival level

(Source: Fawkes, 2007)

Operational level

Infrastructure

In the commercial space tourism industry, all future areas of infrastructure such as space vehicles, spaceports and other technical practicalities should be developed and designed with sustainability in mind. There are already good examples of such actions, for example, in Spaceport America in New Mexico (USA). Here, the 'Gateway to Space' facility has been built with respect for its ancient surroundings while embracing the future through energy efficiency and sustainability. This particular Virgin Galactic spaceport is located mostly underground, powered by renewable energy and is a carbon-negative energy producer (Spaceport America, 2019a).

Green design

Green design is a component of sustainable development with a long historical pedigree. It used to be the norm that utility items tended to be made locally. The ultimate challenge of the 21st century has been to minimise the adverse impact of all processes on the environment and the emphasis has been on products that give maximum human benefit with minimal use of materials and energy (Fuad-Luke, 2010). Today's designers have been able to experiment with more sustainable technological tools by reusing recycled materials and examining alternative systems of production. Redesigning design can itself be seen as a step towards a responsible future (Krippendorf, 1994).

Green design in space tourism could include everything from space suits to spaceport ground facilities and the design of other operational functions. The design of space vehicles, in particular, needs holistic analysis on emissions and should always aim for a 'cradle to grave' life design. These products will determine not only the energy and material inputs, but also the associated environmental impacts.

Carbon offsetting

Since the millennium and especially after the IPCC (2018) report on climate change, there has been a noticeable change in tourist behaviours towards flying and the entire aviation industry. The younger generation in particular has started to optimise their carbon footprint and has demonstrated a preference for alternative ways of travelling, such as trains. Airlines, tour operators and concerned travellers have started to turn to carbon offsetting to mitigate the impact of air transport. The idea is based on 'neutralising' the atmospheric consequences of a holiday by paying more than the ticket price, with the extra invested in minimising emissions in the carbon markets. Carbon offsetting sets the pace of this transition by reducing the financial and political incentives for change and enabling an environment for broader emissions reduction and social and technological innovation. A tourist may ease the stress their ecological conscience causes by selecting an alternative option, such as a sustainable service or payment (Broderick, 2009).

Environmentally conscious tourists can calculate the carbon emission from their activity and offset that by, for example, paying a carbon offset company. Similar compensation schemes should be set either voluntarily or by the policymaking process in the environmental statements to the space tourism industry to reduce the consequences of direct and indirect impact on the environment (Cooper *et al.*, 2008). Many airlines and tour operators increasingly offer their clients the possibility to compensate for the climate impact of air travel by joining voluntary programmes. National- and international-level policy changes on fuel taxes and emissions charges are still rare, although Sweden introduced a Swedish aviation tax in 2018. This means that airlines and air charters with flights from Sweden will need to make a declaration and pay tax, the rate depending on the passengers' final destination (FCC Aviation, 2019).

Sustainability can be more complex when it concerns fuel. Hydrogen is often held up as the ultimate low polluting, sustainable fuel, but this is a gross simplification as hydrogen is only an energy vector, not a source of energy, and is usually generated using power made from fossil fuels. Hydrogen produced from fossil fuels without carbon capture and storage has an equivalent carbon footprint to burning gasoline. As it will take time to develop new green natural resources, voluntary carbon offsetting could provide a useful tool for space companies as one component of their actions towards sustainable tourism in the future. In a similar manner to the current sustainable schemes by airlines, space tourists could also contribute their share and pay more for a 'sustainably conscious' ticket. The price of offsetting a suborbital trip is likely to be approximately only equivalent to a flight from London to Singapore and offsetting a trip to the International Space Station using Soyuz technology would cost approximately £2000 (Fawkes, 2007).

However, it has been noted that in situations involving collective responsibility, the polluter who pays the contribution appears particularly powerless, as they are not able to determine the entity that should bear the costs nor ensure a just process by which compensation can be secured. In the space sector it would in all likelihood be very difficult to reach consensus on the comprehensive application of the polluter-pays principle in outer space and the ensuing channelling of all liability to the operators of environmentally harmful activities (Viikari, 2007).

Critical claims over commercial space travel becoming a significant environmental burden may be absolutely correct in the short term, but not over a longer time period. Clean solar energy could in theory eliminate the environmental impact of space travel, and even make it 'carbon neutral'. It could be a serious error to prevent the growth of space tourism in order to avoid its initial, minor environmental impact, since this would prevent a range of future benefits, including the supply of new carbon-neutral and low-cost solar power from space (Collins & Autino, 2010).

A question to consider carefully at this point is whether it is responsible for humankind to progress space tourism in the hope that these futuristic projects just might someday work out, while under the current climate change crisis (Spector, 2020). There could also be alternative ways of achieving space-based solar power and other benefits rather than being dependent on the development phases of space tourism, such as more concentrated governmental investments and public–private partnerships bringing together scientists and professionals rather than tourists.

Compensation schemes

The commercial space tourism industry could in future follow similar compensation schemes to those already present in the aviation and tourism industries. The means of compensation has differed within organisations. For example, British Airways (2019) has partnered with a renewable fuels company to design a series of waste plants that convert household waste into renewable jet fuel to power the fleet.

Lately, planting new trees has become a global trend as climate change research has presented it to be the most effective solution so far to climate change. Trees remove CO_2 from the atmosphere by photosynthesis, releasing oxygen and part of the CO_2 through respiration and retaining a reservoir of carbon dioxide in organic form. It has been calculated that planting 1.2 trillion trees could cancel out a decade of CO_2 emissions (McGrath, 2019). Countries such as Australia plan to plant trees as part of their effort to meet the country's Paris Agreement climate targets (Yale Environment, 2019).

The effects of compensation schemes primarily target the prevention of environmental deterioration caused by climate change. However, there are some critical questions to consider:

- How does flying with compensation compare from an environmental perspective to the option of not flying?
- Will increased frequencies and new destinations such as space attract people who would not otherwise have flown, but because of the possibility to compensate, choose to do so?
- Will different compensation schemes start to offer 'the best deal' options with relatively low costs compared to rival schemes?
- Will governments act to implement stringent public policies if voluntary actions become standard?

Carbon offsetting suffers somewhat from a low degree of transparency, and a host of other issues, placing the customer in a position where they may not necessarily receive a clear understanding of the impact of their individual donation as it goes into a larger pool and not specifically towards the flight in question. For a customer, it may be simpler to understand the consequences of practical developments such as airlines renewing their fleet with aircraft that use next-generation fuels, as clear figures of the decrease in total emissions can be calculated. It appears as if many environmentally certified projects and schemes aiming to reduce the amount of CO_2 in the atmosphere support projects that are not local to the customers and their air quality. One instance is the Nordic airline Finnair supporting an emissions reduction project in Mozambique, in which the real impact may seem a bit unclear, as with other global social charity schemes. Planting more trees does in theory sound like a reasonable option, but it should be noted that climate change has also had an influence on the increase in fires in many areas, as seen in the Amazon, Californian and Australian fire disasters in 2019. During the crisis alone, Australia emitted half of the country's annual CO_2 emissions (Davidson, 2019).

For example, Finnair (2020) offers the option to compensate by buying biofuel, produced from used cooking oil, which is said to reduce emissions by 60–80% compared to the same amount of fossil fuel, with the explanation that aviation fuel can contain a few percent of biofuel in its maximum of 50% biocomponent. However, as there are a limited number of airports for refuelling, the option is only valid for certain routes. It remains to be seen how such compensation schemes would work with commercial space operators. The operational style of compensation models that support global projects and consider the atmosphere a global matter are already familiar through the actions of the airline industry. A further question appears around whether pioneer 'space jump' tourists would also be willing to support sustainable planning that furthers the space tourism product palette, such as travel to the Moon, or would they just be interested in compensating for their personal footprint on the 'jump' they take?

Cultural level

For the wider public, outer space and sustainability became visible for the first time when Apollo 17's photograph of the Earth beyond the biosphere was taken in 1972. This image became a symbolic icon of environmentalism, presenting the Earth as a fragile object upon which human actions have a great influence (Spector *et al.*, 2017). Fawkes (2007) claims that, without doubt, space travel has helped to catalyse the environmental movement, raising awareness and enhancing participation in global environmental movements such as Earth Day.

Seeing the Earth from space has changed some astronauts' worldview and created a feeling of more connection to the people on the planet than ever before. 'The thing that grew in me over these flights was a real motion and desire... to not just enjoy these sights and take pictures, but to make it matter' (Kathy Sullivan [1984], the first American woman to perform a spacewalk). Some former astronauts are even working together on a project to use their experiences to help others adopt more sustainable lifestyles (Drake, 2018).

The impact of some global weather events, such as the Hadley Cell pattern of hot air, can be witnessed from space, reminding observers that the atmosphere is the only shield protecting humans from cold temperatures and radiation from outer space (Vince, 2016). In this light, it would be very beneficial for climate change projects to receive more funding in the future from the wealthy influential elite who have been the first pioneering tourist group in space and have witnessed the Earth from space. This would help speed up practical climate change actions on Earth. If Earth's atmosphere is destroyed, there is, as yet, no real alternative in the universe to escape the consequences, and as humans are genetically connected to Earth, this could ultimately lead to the extinction of the human race.

In past decades, there has been an expansion of the global long-haul travel megatrend among younger generations, who have been trekking around the globe and sharing their experiences on worldwide social media platforms such as Instagram. Many have already backpacked to destinations difficult to reach by the traditional methods of mass tourism and exposed themselves to genuine risks and dangers. One must not forget, though, that privileged youth trekking around the world to experience nature has also had a negative impact on remote environments and the global climate as the number of travellers has increased. When space tourism reaches a wider market, these same people could potentially also want to experience space and share their feelings with their social media followers and other peers.

Children and teenagers have always found the subject of space and space travel fascinating, as evidenced by the popularity of a number of space-themed television series such as *Star Trek*, and science fiction films, the most recent addition being *Ad Astra* (2019), which has so far displayed one of the most realistic depictions of space tourism in film. Many organisations and institutions around the world offer space-related educational programmes and scientific education. However, there is an existing social division between 'rich' and 'poor' countries in the variety of available affordable space education options. In the worst-case scenario this could lead to a threat to the successful continuation of civilisation as some of the potential audience cannot be reached.

Interest in experiencing future space travel, if it becomes similar to aviation now, or even working in space, is likely to be much greater among young people who have been exposed to space-related education, virtual videos and space games. The start of low-cost passenger space travel services holds a unique promise for education in fields related to space travel. The opening up of the space frontier may have profound effects on culture as well. In a similar manner to the Renaissance in Europe and the industrialisation of the 18th century, which led to the beginning of the travel industry, it could evoke a new sense of global optimism and hope for the future (Collins & Autino, 2010).

Economic level

Bank of America Merrill Lynch has projected that the commercial space sector will be worth $2.7 trillion by 2045 (Saigol, 2019), with the vast majority of the economic impact coming from satellites. However, as a part of this lucrative space-related sector, space tourism could become a new driver for global development, and ultimately for human development beyond our planet and solar system. Throughout economic history, technical innovations have created new industries in the developed world which have replaced old ones that have then moved to developing countries, creating a cycle of need for creating new industries to create new wealth and employment (Fawkes, 2007). Initially, terrestrial launch sites for commercial spacecrafts will provide a new opportunity for the economic development of the local area and community. They could bring local benefits in terms of generating new jobs, income and revenue, and assist in building the local skills and culture and the area's ecology.

Commercial space travel and its numerous spin-off activities also have the potential to escape the limitations of consumerism, which governments in wealthy countries have encouraged in recent decades in order to stimulate economic growth. This has resulted in excess consumption, causing unnecessary environmental damage and reducing, rather than increasing, satisfaction. First World citizens have increasingly been enmeshed in a culturally impoverished consumer lifestyle which has reduced social capital, social cohesion and happiness, while also damaging the environment. By contrast, expenditure on the unique experience of space tourism promises to play a more positive role in the economy and society, enriching customers culturally without requiring mass production of consumer goods and the corresponding pollution. As such, it could be a harbinger of a future open world economy (Collins & Autino, 2010). On the other hand, one could question whether space tourism is actually the ultimate consumerist act which takes money away from more concrete project initiatives, such as alleviating global poverty or hunger (Spector, 2020). Perhaps such projects would otherwise receive funding from the same individuals if it had not been spent on space tourism.

Economic development in space could also contribute to solving the world's environmental problems. Firstly, the reduction in the cost of space travel would reduce the cost of environment-monitoring satellites, thereby improving climate research and environmental policymaking.

Secondly, the delivery of continuous solar-generated power from space to Earth could supply clean, low-cost energy on a large scale, which is a prerequisite for the economic development of poorer countries, while avoiding damaging pollution. The realisation of such power is dependent on much lower launch costs, which, presumably, only the development of a passenger space travel industry could achieve. The development of orbital tourism could, therefore, provide the key to realising such a source of power economically. The use of solar-powered satellites could even reduce the severity of hurricanes or improve severe snow conditions, stabilising the climate with a positive feedback cycle. For example, satellite power stations may be the only practical means of selectively melting snow over areas of thousands of square kilometres, even in the event of terrestrial energy shortages (Collins & Autino, 2010).

Resource level

Earth has sustained life by providing accessible resources to humans and a multitude of living organisms, creating a symbiotic relationship between the utilisation of resources and civilisations. The human race has survived by adapting to the environment and now technology has opened up access to other planets to seek and even utilise their accessible natural resources. While the quantity of different natural resources in space, including metals and minerals, is vast, providing future business opportunities beyond imagination, NASA's technology development project In Situ Resource Utilisation (ISRU) has so far focused on three major types of space resources: regolith, atmosphere and solar energy, which are important in ensuring suitable future living conditions for humans in space (NASA, 2019j).

For decades, travelling to space may have appeared to be a waste of the Earth's resources, as accessing space and developing the entire space industry have demanded vast quantities of natural resources, without bringing similar physical resources back to our planet. This all changed after the millennium when rapid technical developments created a need for satellite assistance for many daily and necessary activities around the globe. Various routine activities, such as using navigation while driving and computer and smartphone usage, are nowadays dependent on the satellite network operations orbiting the Earth.

As humans around the globe have become heavily dependent on such purpose-built space resources, this creates a valid question over power and ownership. The individuals, companies and governments involved in space-related activities are those already in positions of comparative power and wealth and as spaceflight technologies rapidly progress, the wealth continues to be concentrated within a limited number of groups (Spector & Higham, 2019a).

The exploitation of space resources creates ethical concerns, especially as space legislation is still under development: who actually has the right to take over space resources and exploit the available resources commercially? Even though the creation of commercial space tourism does not mean there will be traditional industrial mining for metal and mineral digging, the industry will require physical landing sites, such as the Moon's surface, and the exploitation of the natural resources of space to accommodate the physical needs of the future tourists since not all necessities can be delivered from Earth.

Multi-scholarly research has already indicated that virgin stocks of several metals will soon be inadequate to sustain the modern developed world's quality of life for all Earth's people using contemporary technology. Even though there has been an increase in recycling and the substitution of materials, there is the ultimate question of resource availability for the basic raw materials of industrialised society in the future (Fawkes, 2007). The possibility of exploiting space's vast untouched resources has become a tempting alternative, but also creates concerns that utilising the abundant resources of the solar system could lead to a catastrophe on Earth by fuelling a consumption boom (Cockell, 2007). It is essential to implement global sustainable development policies among all active actors, such as governments and private companies, to avoid such an unsustainable future scenario.

The space industry's growth to a large scale could offer a longer-term possibility of decoupling economic growth from the limits of the terrestrial environment, and, in the extreme scenario of only using space resources in the future, could protect the Earth's environment, while enabling sufficient economic growth to preserve civilised society. This kind of action would create the ideal sustainable environment on Earth, but whether this would be a morally sound path to take is another question – does the human race have the right to rescue its own planet's resources by colonising areas for resource utilisation in space, or does space also have sustainability rights of its own that should be respected?

The space industry is likely to develop in line with the growth of orbital travel, stimulating the advancement of other businesses in space, in industries varying from orbiting hotel maintenance to space manufacturing using asteroid minerals and benefiting from the advantages of space such as weightlessness and high vacuum. Progress in using solar power and helium 3 could supply a significant share of the terrestrial energy market, leaving more industry to operate outside the Earth's ecological system, thereby protecting the Earth's scarce natural resources (Collins & Autino, 2010; Fawkes, 2007).

Survival level

> Our only chance of long-term survival is not to remain inward looking on planet Earth, but to spread out to space.
> Stephen Hawking, 2010

The extinction of dinosaurs 65 million years ago from a presumed asteroid or meteor impact already demonstrates how easily a living history of 160 million years can be wiped out from planet Earth when the living conditions suddenly turn hostile. The Earth's position in the solar system means that outer space meteors and asteroids constantly present a potential threat of collision with our planet with the potential to have serious implications on the ecosystem. As with the dinosaurs, the human race also needs suitable living conditions in order to survive.

Over the years, there has been much speculation about how the world could end for humankind, including big catastrophes such as global disease pandemics, nuclear wars, super volcanoes, mega tsunamis, gamma-ray bursts and asteroid impacts – most would be capable of changing living conditions on the planet. Lately, the realisation of shortages of natural resources, such as clean water, and the impact of climate change threatening all life on Earth, have caused major uncertainty for the continuity of the human race under such a living environment. The tourism industry has also contributed to climate change, particularly through the creation of carbon emissions from tourism activities, with transportation being one of the most influential contributors (Cooper et al., 2008).

The only practical solution to sustain the future of human life, as life on Earth faces so many potential threats, is to spread into space and establish future space colonies as a human insurance policy. However, it has also been speculated that there is only a limited window of opportunity to develop such space colonies, with the assistance of lower-cost space access due to space tourism services, before the Earth's resource problems become too extreme to provide the resources available for space travel (Gott, 2002).

Conclusion

Sustainable planning for future space tourism may at first appear to be an impossible task to accomplish, due to the elements commonly associated with both tourism and space travel, as trendy practices in tourism concentrate on keeping nature and destinations as untouched as possible (Mowforth & Munt, 2015). Thus, destination life cycle models such as TALC (Butler, 1980) and Plog (1974) provide a necessary tool for guiding the planning process for sustainable destination development in new tourism environments.

Many practical actions and other sustainable developments can be linked to future space tourism at different levels, providing enhanced cultural education for the younger generation and fostering the urge to protect the Earth to ultimately save Homo sapiens (Spector et al., 2017). The creation of the space tourism industry also facilitates a reduction in the future cost of reaching space and its vast resources, possibly in the future enabling some resources such as solar power to be increasingly utilised on the Earth, thereby saving the planet's resources.

Furthermore, technological innovations in the operation of space tourism may produce alternative energy sources and practices for Earth-bound aviation and other ground transport. The risks caused by creating a 'non-sustainable' image of the future space tourism industry should be mitigated from the beginning by resolutely introducing sustainability into different operations of the space tourism industry and also clarifying that the space environment itself will have an important role in sustainability.

5 Space Tourism and Society

> We humans have to go to space if we are going to continue to have a thriving civilization. We have become big as population, as a species and this planet is relatively small. We are in the process of destroying this planet.
> Jeff Bezos, 2019

Introduction

It is important to consider some of the elements of social sustainability, in which the core idea involves identifying and managing business impacts, both positive and negative, on people (UN, 2019c). The emergence of the new space economy, including space tourism, will have a local impact in a variety of ways in many areas of the planet, such as economic opportunities and environmental threats, and potentially even inform the way future societies function.

This chapter first explores the emergence of space tourism in the 'last remains' of the Anthropocene, in which human actions have already had major consequences on the Earth's climate and environment. Secondly, the chapter looks at some elements impacting change, which cover many activities, and include, for example, activities formerly practiced only by professionals, but which have become increasingly shared with holidaymakers. Considerations of tourism degrowth and social fairness will also be explored. Thirdly, some ethical considerations of space tourism, including corporate ethics and space commercialisation, are investigated. Lastly, the impact of spaceports on local communities is examined. The United Kingdom is featured in a case study, and some global spaceports of significance are introduced.

The Anthropocene: The Human Era

> Space tourism provides a germane context for conceptualising the ongoing debates regarding the extent to which Anthropos – humankind as an undifferenced geological force – is responsible for the impacts that have culminated in the Anthropocene.
> Spector and Higham, 2019a

To understand the changes that have taken place on the planet, it is first wise to consider the historical context. The geological period of

Holocene started 11,700 years ago, at the end of the last glacial period of the Pleistocene and provided a relatively warm and stable climate as well as fertile environments for plantation. During this period, human habitation expanded significantly northward. The Holocene also accelerated the development of human societies through the processes of agriculture, control of water, industrialisation and urbanisation – all contributors to modern-day environmental changes (Spector & Higham, 2019a).

The concept of the era of the Anthropocene was brought to attention by the Nobel Prize-winning atmospheric chemist Paul Crutzen in 2000 (though it was created in the 1980s by Eugene Stoermer), referring to the actual impact and evidence of the effects of human activity on the planet Earth (Crutzen & Stoermer, 2000). The message was that the Holocene had ended and a new epoch, the era of the Anthropocene, has begun. The Anthropocene is still debated, and the concept has also been criticised for the abstract conceptualisation of humankind that some proponents appear to adopt. There is also an ongoing debate, especially between geologists and environmentalists, as to whether humans have permanently changed the planet, as there is scientific evidence of mass extinction of animals and plants, polluted oceans and atmospheric alternations (Strömberg, 2013). There is, in general, however, a growing acceptance from many different fields that we have entered a new, human-induced geological age, and the concept of the Anthropocene has also become prominent in the tourism literature (Spector & Higham, 2019a).

The origins of the Anthropocene date back to the first steps of the Industrial Revolution, particularly because the industrial use of fossil fuels has resulted in a human contribution to pollution, over-exploitation of resources, population growth and species extinction (Moore, 2015). Viikari (2007) states that the anthropocentric slant of the space sector derives from the historical phenomenon of industrial development as all human activities in outer space have been made possible by the achievements of technology. The ideology prevailing within the modern space sector emanates from the fundamental concepts which this industrial tradition entails, above all the myth of unlimited industrial development. In the space sector, the shortage of resources to be used in outer space cannot be ignored and there are high hopes that space's resources will provide a solution to the increasing scarcity of resources found on Earth. The faster the technology improves, the more feasible the utilisation of outer space becomes (Viikari, 2007).

Climate change is the prominent factor in the Anthropocene and examples can be seen in behaviours in Western society: for example, an average United States citizen emits tens to hundreds of times more carbon dioxide than individuals living in less-developed countries. The other 'high emitters' are the 20% of people who fly and are responsible for such anthropogenic emissions, compared to the rest of the world's population, who have never entered an airplane (Anderson & Bows, 2011; Spector & Higham, 2019a). The IPCC (2018) report also raised concerns around global greenhouse gas emission pathways and urged the strengthening of the global response to the threat of climate change in the form of sustainable development within societies (IPCC, 2019).

Climate change can also be considered sociogenic rather that anthropogenic and referring to anthropogenic climate change might conceal the social relations that are integral to understanding the phenomenon (Malm & Hornborg, 2014; Moore, 2015). Spector and Higham (2019a) state that the 'Capitalocene' framework can assist social sciences in elucidating the social relations that have created the conditions of the Anthropocene and bring to the fore the role of capital in providing impetus for the issues that have led to the era. One should not, however, replace the Anthropocene with the Capitalocene, but use it to provide useful insights (Moore, 2014). The Capitalocene itself focuses on issues of the social systems under which usage occurs and the historic developments (such as those made during the Industrial Revolution) that have led to contemporary socio-environmental issues (Spector & Higham, 2019a).

Changes in society

New forms of organisation and production, developments in transport and mobility and the rise of information technologies have resulted in an intense time–space compression, allowing new experiences of space and time. People on Earth may be considered not only space travellers, but also time travellers. This metaphor reflects on the fact that the travel industry is currently taking the next step towards the future, especially as First World citizens have increasingly become trapped in a culturally impoverished consumer lifestyle, which has reduced social capital, social cohesion and happiness, while damaging the environment (Bell, 1997; Cole, 2015; Harvey, 1989).

Travel is the most obvious luxury product experience, offering an opportunity to spend time in unique locations, and commercial space tourism is the next-generation luxury experience to provide affluent travellers with unique experiences that are difficult for the average person to replicate (Wittig et al., 2017). The development of outer space will have significant implications for Earth's inhabitants, yet only a small section of individuals, companies and governments are currently involved in the process (Spector & Higham, 2019a). The question of inequality has already been noted in recent UN reports on climate change and poverty. Many countries have ignored the scientific warnings about climate change and have not followed targeted emissions cuts, causing global critique. It has even been forecasted that the world is soon to face a 'climate apartheid', in which the rich can escape the consequences of global warming by emigrating to space, for example, leaving the poor to suffer from the impact (Alston, 2019).

Trends in adventure travel have also blurred the boundaries between adventurous activities and tourism (Beedie & Hudson, 2003). Mountain climbing, previously practised only by the experienced elite, now shares popularity with holidaymakers attending a guided and safety-checked experience. Similarly, space has previously been accessed only by trained astronauts and scientists observing planetary movements, not by ordinary tourists. The desire of tourists to have more unique and challenging experiences has been one of the driving forces behind the demand for public space travel, pushing for the development of a new adventure tourism sector entering a virtually untouched new space.

The recent megatrend of environmental protection, globally promoted by Greta Thunberg and other young national influencers, can no longer be ignored by any industry. Concerns about the impact of climate change will have an enormous influence on the consumer and travel decisions made by 'Generation Greta'. In order to succeed in the future, no industry can afford to neglect the environmental impact of their own actions; responsible environmental strategies for their business are a requirement.

Tourism degrowth and social fairness

Social sustainability also relates to globalisation, relationships of power and uneven and unequal development. In many bodies of research there have been various categorisations of countries according to their wealth and social well-being. The new forms of tourism, prefixed for example with 'sustainable' and 'eco', have attempted to signal that there has been an attempt to move away from 'plain old tourism' with its negative impacts and let the tourist believe that the 'old problems' have now been overcome sensibly. However, despite attempts at sustainability, new tourism developments may have brought additional negative impacts to communities in the form of environmental damage, congestion and noise, for instance. The community may have been better off without the tourism activity (Cooper *et al.*, 2008).

There have been decades of concentrated effort to achieve sustainable development, but socioecological conflicts and inequality have rarely reversed – in fact they have increased in many places. This realisation has resulted in new emerging development and theoretical approaches to tourism degrowth to provide more equity for host communities, by offering, for example, a lower carbon future by looking at alternatives to the classic models of tourism development. The concept of tourism degrowth offers some understanding of both social theory and social movement, which has emerged especially within the context of global crises (Fletcher *et al.*, 2019).

In many of Earth's tourist destinations, there has been a paradigm shift from 'tourism growth' to 'tourism degrowth' (Milano *et al.*, 2019). 'Touristification' of cities and natural places has been increasingly met by the discontent of local communities deprived of their environment, causing 'overtourism' to be a real issue with various challenges. Thus, there needs to be rethinking and remaking of the tourism industry to bring tourism and degrowth closer together (Fletcher *et al.*, 2019). Overtourism has become a rapidly evolving contemporary tourism phenomenon, often an uncontrolled and unplanned occurrence, resulting in social movements in the form of protests and activists' campaigns (Milano *et al.*, 2019).

Of course, 'destination space' is hardly in its pioneering stage and facing socioecological conflicts due to overtourism; its high costs limit the clientele. However, the tourism degrowth discussion will certainly be valid in the future to ensure sustainable future planning for space tourism. Overtourism is a structural response to the ravages of capitalist development and debates concerning overtourism offer valuable opportunities to re-politicise the discussion of tourism development generally in order to facilitate genuine sustainable tourism (Fletcher *et al.*, 2019).

The power division in relationships has also traditionally ranged from the power of First World countries in contrast to Third World countries, to the power wielded by international donors, from the World Bank to the European Union, the power of local elites in contrast to local populations in tourist destination communities and the power invested in tourists themselves. A distinction can also be made between individuals, as society consists of people from different backgrounds, some from the wealthy elite while others face daily struggles in fulfilling their basic daily needs like food and accommodation. There have always been countries and people in more powerful positions than others, which has resulted in questions about equal fairness, especially if the impact, such as polluted air, needs to be suffered mutually, despite only the wealthy individually benefiting from it (Mowforth & Munt, 1998).

Postmodernism arguments related to tourism have often led to the conclusion that just as an increasingly globalised world has resulted in a global economy, the same process has, at the same time, also resulted in the emergence of a global culture characterised by the same global status symbols and material products (Harvey, 1989). This phenomenon has become especially visible since the emergence of social media, as increasing numbers of younger Western travellers, who capture similar travel photos in the same destinations and post these on the internet for others to see, join the mass 'status' travellers. The First World perspective on travel has become more dominant than ever, defining the most appropriate way of holidaying, with the prospects and problems of tourism calculated also from that perspective.

Postmodernism can also be linked to the emergence and growth of the so-called new middle classes, forming key social groups in initiating, transmitting and translating new cultural processes and travel consumption patterns. With the speed of the world's cultural and economic change, the social groups attempt to identify and indicate to others their position in the new social and cultural order. Globalisation has captured the idea of a shrinking and more equal world, with the result that countries have become increasingly interdependent, and in tourism the rise of a middle class has created new types of travellers (Mowforth & Munt, 1998). In relation to space tourism, daydreaming of travelling to space has traditionally presented as an equal escape attempt in which people participate by jetting, physically or not, away from social monotony and from themselves (Cohen & Taylor, 1992). However, for the foreseeable future, physical escapism is likely to involve those already in the most powerful economic, social, cultural and political positions, deepening the issues related to responsibilities and fairness (Höckert, 2015; Ormond & Dickens, 2019).

In space tourism, the division between different social groups will be very clear in the pioneering stage. As the estimated ticket cost is $250,000 per 'space jump', it is obvious that only the wealthy elite is able to access this new sector. The rest of society must save for years or place their hopes on a space tourism 'lottery' type of goodwill gesture, in which space tourism companies provide cheaper or even free tickets to those with influential marketing backgrounds, such as Instagram bloggers, media representatives, educational professionals or other such groups.

It remains to be seen how most travellers will react to space tourism, where access will be determined by financial capacity and not yet experienced authentically through other channels. Of course, space tourism will be considered luxury tourism, where the context of inequality is already present. It costs tens of thousands of dollars to be able to climb Mount Everest. However, most travellers from Western countries have been able to enjoy similar travel experiences to the wealthy elite, simply by selecting differently, such as staying in a budget hotel, using public transportation and self-catering instead of eating out. Access to the physical destination, even to Mount Everest, has remained equal to the wealthy elite.

With 'destination space' similar budget optimisation will not be possible in the pioneering stage, as space cannot be accessed without a specially designed space vehicle. This may result in strong class divisions, such as in the case of the Grand Tour to Europe hundreds of years ago. This raises the question of how moral it is to launch a luxury tourism activity only for the pleasure of a few, which will be difficult for most to replicate. Will this direction towards elite exclusive tourism create a divide in the Western world similar to that between the First World and the Third World? In this case, it could shake the relatively stable behaviours of Western society, and even lead to acts of anarchy, as younger Western generations in particular have become accustomed to equality and possibilities for all. One might also question what right Western countries have to establish utopic elite forms of tourism from the rest of the world's perspective, where there are still enormous social issues related to poverty and hunger; such development money could instead be spent on first solving the world's problems (Spector, 2020).

However, there have already been some signs that consideration has been given to fairness issues by the affluent behind space tourism operations. For example, in February 2020 the world's richest man Jeff Bezos, behind the Blue Origin space tourism company, unveiled a fund of $10 billion to help climate change scientists and activists address climate change, which he described as 'the biggest threat to our planet' (Weise, 2020). Such research funding could lead into more hegemony in the development of global sustainable politics to enhance more equal rights and opportunities distinct from the current global power jigsaw.

Ethical Considerations

The ethics and philosophy of space exploitation arise from activities that are congruent with the last remaining imperial and mercantile frontier (Duval & Hall, 2015). Human interest in space has previously related to scientific values, but the development of the commercial space tourism sector now introduces space as a new tourism destination to be experienced physically. Thinking in terms of ethics, in order to minimise mistakes similar to those made here on Earth, the exploration of other planets should continue to be based on research, rather than purely as a means to ensure that there is a place for human beings to replicate the Earth's living conditions.

Alternative ethical approaches to the economic benefits and environmental impact of tourism raise issues related to responsibilities and fairness (Höckert, 2015). As with many periods in history, human actions have tended to prioritise economic growth, even with the destruction of land, wildlife and local populations. A question to be raised is do humans actually have any moral right to take over other planets in the universe?

Mixed feelings are present in different scientific views: some scientists stress that humans have the moral duty to treat potentially alien creatures alike, even if they are only rudimentary microbes and, vice versa, others argue that the ethical angle to actions is not necessary until proof of life is actually discovered. However, one way to begin an ethically sound future for potentially intrusive activities could be to begin mining from asteroids, as they do not likely present any form of life. In this way, humans prioritising their own needs would not force any other life form to disappear (Peltoniemi, 2018).

The questions surrounding space ethics have resulted in speculation on the future of the space tourism industry and possible spin-off impacts of space colonisation. McFadden's (2018) investigations for the article 'SpaceX and the Ethics of Space Travel' raised some valid key issues to be taken into consideration. The findings revealed that there are two sets of ethical concerns related to the expansion of space travel:

(1) Bioethical concerns.
(2) Political concerns both among the nations of Earth and between Earth and those that venture off-planet.

Bioethical concerns relate to the contamination of space and have been one of the major considerations since humans set foot, either physically or mechanically, in space. The issue becomes especially real as there are a number of models and physical evidence seen here on Earth of how an invading species can wreak havoc on a planet's environment. For example, on the remote Galapagos Islands, where visiting boats brought invasive species such as ants, a domino effect leading to the disappearance or a reduction in some of the original species has occurred (McFadden, 2018).

As it is impossible to completely sterilise the materials and personnel entering space, regulatory bodies, such as NASA's Office of Planetary Protection and the International Council of Science's Committee on Space Research, deal with the ethical concerns of introducing Earth-born organisms to alien worlds. Governmental involvement ensures that research on the technological sterilisation solutions can be supported in the long term. NASA's Office of Planetary Protection promotes the responsible exploration of the solar system by developing and implementing efforts that protect science, explored environments and the Earth. Ultimately, the objective is to support the scientific study of chemical evolution and the origins of life in the solar system (NASA, 2019k).

Political concerns may be divided across two levels. Firstly, there is the complication of who has the legal jurisdiction over what happens in space now that more than scientific goals are being pursued. Companies are already trying to provide real estate on the Moon, or even Mars, to private individuals and, similarly to companies that sell 'naming of the stars' certificates, there is no real legality to such activity, although these kinds of transactions may be subject to possible future national claims over ownership. Once private space tourism companies such as SpaceX have the capability to operate tourists trips to the Moon or Mars, they are likely to need a base to operate from and this raises the question of whether a resort area, for example, would fall under the legal jurisdiction of the home country of SpaceX or could it be a no nation's land, legislated by SpaceX? A mutual or a shared platform ownership could be one solution to solve the issue of legislated ownership in space. The International Space Station presents a good operational sample of such ownership as it follows shared agreements on the operation and findings of the experiments between Russia, the United States, Europe, Japan and Canada (McFadden, 2018).

Secondly, there is also the question of equal possibilities regarding the areas in space. In the future, when space exploration is in full operation, would the nations without access to space have any right or claim on any space-related actions and legislations? For example, would it be possible to prohibit the access of a certain nationality to certain extraterrestrial areas if the person is not from a 'space nation' country, and would there be regulations similar to visas for such tourist requests? Another ethics-related political concern around sending people into space is that of political legitimacy itself. Questions arise about who should actually hold the governing rights over the land in space and what would the relationship be between the yet unknown regulatory body on Earth and those living or visiting off-Earth. As the first pioneers in space are risking their lives trying to start a human society on another planet, should they be allowed to have some compensation for their efforts and even gain a certain level of independency or would such colonies always be dependent on the rules and legislation originating from the Earth (McFadden, 2018)?

The way of life in space will differ significantly from that of the governing body and laws needed may be completely different and raise discussions around what can be regarded appropriate (Viikari, 2007). For example, in the case of crime, would the limited resources and concern for survival have an impact on the severity of punishments, perhaps minimising them? And if a colony suddenly started rebelling against the country of origin, how far would goodwill go in providing resources from Earth to maintain its survival (McFadden, 2018)?

Corporate ethics

Marsh (2006) explored elements of organisational ethics in the article 'Ethical and Medical Dilemmas of Space Tourism'. Firstly, some high-profile large industries of today, such as tobacco companies, can readily serve as examples of why there are ethical concerns about exporting the business models of the late 20th century from Earth to space. The company's duty to fulfil human needs in a responsible way is valued highly in the current business environment and competition for customers has left consumers with the power to select a company following the social laws and morals of the purchasing individual. Corporations not following the general ethics of society have made themselves liable to lawsuits. Without an ethical orientation to the conduct of one's business, people can be made to suffer extreme harm as business decisions can have the power to impact on individuals' lives.

Ethical issues can also arise as a result of the conduct of the business or its management, or even just because of worries about such potential conduct. For example, loose comments made by influential people on social media platforms, such as Twitter, may be enough to suddenly foster more concerns around ethical issues – even if the statements are originally meant to be nothing more than harmless rhetoric. However, ethical issues are seldom equated with legal issues and will most likely not be addressed in a legal forum (Marsh, 2006).

Consumers' conscious participation or neglect in influencing ethical standards for the commercialisation of space can shape the character of space commerce and those living in space in the future for a long time. As humans begin working, living and establishing trade routes and businesses in space, space will for all time be characterised by the impression humans place on its blank pages (Bluth, 1979). Humanity has yet to face a challenge, opportunity and responsibility of this magnitude. One of the initial problems in considering ethical standards for space commercialisation in general is deciding whose ethical standards are the 'right ones' to follow. Different ethics are present within different cultures and it may be a complex decision to decide which ethical standards have added the most value to cultures and customs, to be valuable enough to be copied as moral guidance in the space environment. Marsh (2006) points out that selecting these sort of 'Ten Commandments' could have been easier some decades ago, copying the ethics of Western society, rather than in the current global environment.

Space commercialisation

Some professionals generally supportive of space commerce and colonisation have pointed out some potential risks awaiting us in unchecked and unsupervised space development. Some of the concerns date back to the 1970s, when space commercialisation simply meant satellites, suggesting that large-scale colonisation should be postponed until such a time as social and political conditions reach the prerequisite state of sophistication – if ever. As the private space industry, such as SpaceX, has started to take a new lead in space exploration, governments are no longer the sole players in space development.

The free-enterprise system will most likely be applied in outer-space commerce and its power should not be underestimated because the positive benefits that flow to economic systems and societies supporting the development of free-enterprise systems have usually outweighed the abuses and problems. Marsh (2006) states that the fundamental questions in the commercialisation of space can be summarised as follows:

- What will the voyage into space look like?
- What will humans' space commerce look like?
- Will space commerce directly resemble the business and management models exported from Earth?
- Will businesses with the right economic power export operations to low-Earth orbit?
- Will the Earth's terrestrial economy be impacted if future commerce expands away from the Earth?
- What is the correct behaviour in commercial activities in space and who decides which operator should have more power in outer space?

The permanent presence of humans in space will affect many existing institutions in terms of economics, science, politics, religion, social relations and psychology (Bluth, 1979). There is also the matter of law and order; for example, in the case of a space colony rebellion, who would be responsible for what, and can space programmes be genuinely committed to peaceful exploration of the space frontier without playing to the dictates of national interests? (Marsh, 2006).

Local Impacts of Space Tourism

Commercial space tourism has the potential to develop and become a new driver for global development, and ultimately for human development beyond our planet and solar system. The terrestrial launch sites for commercial spacecraft, as well as space theme parks and museums, will provide a new opportunity for economic development for local communities. To follow the guidance of social sustainability, the hosting of spaceports or other terrestrial space tourism sites should be beneficial to the local population in terms of generating new jobs, income and revenue and assisting in building local skills, culture and ecology (Cooper *et al.*, 2008).

Traditionally, the multiplier effect is one of the important concepts in tourism planning. The effect is created as money that is brought into an area is re-spent on additional goods and services, encouraging growth in the primary and secondary sectors (Khan *et al.*, 1990). The tourist activity itself determines the most relevant multiplier, whether it is related to factors such as employment, but the net contribution from tourist activities to the local community acts as the main element. Problems for locals may occur if international private industries become involved with no or only a limited intention to involve the local community. In the worst-case scenario, the model for conceptualising space tourism follows the example of the cruise ship industry where spaceports become all-inclusive multifunctional 'spa resort experiences', leaving revenue to the local community almost non-existent and possibly even minimising the tax paid to the host country.

Future off-planet space destinations may also develop into similar 'spa resorts' to those at Earth-based spaceports, dividing the gap between the local colonists and the all-inclusive travelling space tourists, and causing problems if tourists share with the host community the same infrastructure that provides the essentials such as water, oxygen and electricity. In such a scenario, it is possible that tourists would not contribute enough to long-term maintenance. There could also be questions around which legislation, administration or currency these isolated, privately owned space resorts would operate under and, in the case of an emergency, which party would be responsible for actions to provide medical care or law enforcement (Cole, 2015).

Employment generation

The continued growth of civilisation requires the ethical evolution of resources sufficient to ensure health, comfort, education and fair employment for all members of society. In many countries, most local community members do not have economically significant land holdings so local employment opportunities provide the economic basis for social life, general well-being and enabling people to plan for their future activities such as buying a house. Where there is a high level of local unemployment, there is widespread poverty and unhappiness, which is socially damaging, creating further problems for the future. There is a social cost to increased economic efficiency that easily creates unemployment and therefore governments concerned for public welfare should either increase the rate of creation of new industries, such as space tourism, and/or slow the elimination of jobs, at least until the growth of new industries revives, or other desirable countermeasures are introduced. These may include more leisure time, job-sharing and other policies designed to prevent the growth of the 'working poor' – a development which already poses a major threat to Western civilisation (Collins & Autino, 2010).

Future commercial passenger space travel services and space tourism could create employment in vehicle operations and maintenance at spaceports, in orbiting hotels, the companies supplying to these, in services such as health and safety certification, insurance and staff training and a range of other businesses. The growth of such an advanced aerospace technology market could actually be viewed as highly desirable from the local point of view. It has also been speculated that the opening up of near-Earth space to large-scale economic development, based initially on passenger space travel services, promises to create millions of jobs with no obvious limits to future growth. During periods of higher global unemployment, newly developing industries should be given priority when forming employment policies (Collins & Autino, 2010). To continue economic rationalisation and globalisation while not developing space tourism could be self-contradictory and both economically and socially harmful.

Corporate sustainability

Research conducted at the University of Melbourne's sustainability institute pointed out that the space industry itself must first become a sustainable employer to ensure long-term sustainability in other operative actions (Matthews, 2019). There are various methods to assess corporate sustainability, such as the model created by Benn *et al.* (2006) (similar to Fawkes, 2007), introducing six stages for organisation-based sustainability. The stages are:

- Rejection
- Non-responsiveness

- Compliance
- Efficiency
- Strategic proactivity
- Sustaining corporation

Matthews (2019) used the model to assess the sustainability of SpaceX, based on the information on their public website, benchmarking the company with 'compliance' within the sustainability framework. His findings included:

- SpaceX has been innovative in designing ways to travel into space; however, the innovations have not been made for environmental reasons but focused more on the cost of the launches.
- SpaceX relies heavily on government contracts (with NASA), creating further questions for its profitability and capital being raised through loans. The loss of such contracts could have a major impact on the employees of SpaceX.
- The working culture of SpaceX rates poorly with sustainability as many workers have been known to work more than 80 hours a week even while taking mandatory breaks, enhancing a lawsuit settled in 2017. Such behaviour contravened Goal 8 of the UN Sustainable Development Goals, seeking to achieve 'decent work for all'.

The suggestions to attain more corporate sustainability included:

- Involving sustainability principles into emerging space start-ups, so the economic cost of having to correct bad behaviours later can be avoided.
- There will be first-mover advantages on implementing these principles as such actions will increase investor confidence and improve company valuations.
- To ensure long-lasting benefits for the space sector, its defining principle must be sustainability.

(Matthews, 2019)

Case Study: United Kingdom Spaceports

In June 2017, the Queen of England spoke about the development of spaceports in the UK for the first time, mentioning that such ports could function as bases from which both the paying public could be launched into space and suborbital flights could be taken, with a dramatic reduction in flight time between the cities of the world. There is a requirement for much more space research in order for the British government's commitment to becoming an attractive place for commercial space flights to be accomplished (Rincon, 2017). However, some new space legislative measures were introduced just a year later in 2018. To compare the future space strategy within Europe, the Arctic city of Kiruna in Sweden offers examples of space-related operations and competences, providing an existing hub for ground stations for launching satellites – and possibly also a base for European commercial space flight activities in the future (Spaceport Sweden, 2017).

The UK's geography is generally well-suited for launch sites as the northern latitude provides access to specific types of orbits around the Earth. These include polar orbits that are useful for the observation of the Earth. The proximity of the Atlantic Ocean increases safety as the launches can take place over the ocean and avoid populated areas. Currently, many critical and commercial services such as navigation, weather reporting, telecommunications and financial services, and the observation data for these, are provided by satellites. Together with the decreasing cost of launching small satellites, this is driving a global demand for commercial satellite launch services. The government and space industry consider a UK-based launch site to be a key factor required to capture this market and capitalise on the already strong industry in satellite manufacturing and services (Hutton, 2019).

In the UK, the space industry sector has tripled in value since 2000 and year-on-year growth has been five times greater than the wider economy since 1999. Over £415 million was invested by the industry in research and development alone between 2104 and 2015. The government and the space industry have ambitions to capture 10% of the global market by 2030 as the government has estimated that commercial launch demand could be worth over £3.8 billion to the UK economy over the next decade (GovUK, 2018).

Plans are that the spaceports will create jobs in the local area and bring wider economic benefits through, for example, tourism to spaceports, inspiring science and technology careers and the usage of satellite data by businesses and the public sector (Hutton, 2019). Four key priority points have been established to reach this goal, all identified in the Space Growth Partnership Strategy (2018). One of the goals is to develop space launch sites, such as a vertical launch site in the Highlands (Sutherland) and horizontal sites at Cornwall and Glasgow Prestwick Airport (Legislation UK, 2018).

Space tourism options can also be touched on, as the cost of getting to space will become cheaper, making space tourism a real future possibility. There are, as yet, no specific plans to use spaceports to launch people into space in the near future and, in fact, launches proposed to take place at Sutherland are not of a suitable size for space travel. However, Glasgow Prestwick Airport will provide some future opportunities for passenger space flight. In July 2018, the government announced a £2 million fund to help develop horizontal launch sites in Cornwall, Snowdonia and Glasgow Prestwick, which in theory could go towards developing some space tourism facilities. Similar types of funding in the future might prove beneficial because the UK will no longer be involved in European Union programmes due to Brexit, so the emerging prospect of future human flight from British soil as part of a 'national space programme' (similar to France's Space Command) could bring new and powerful enthusiasm to the post-Brexit era, when traditional industries might still be battling with the economic impacts caused by the COVID-19 pandemic.

Space Growth Partnership

The Space Growth Partnership was published by the government on 11 May 2018 to set up a vision for enhanced growth in the space sector over the next decade. The strategy focuses on four sector priorities:

- Earth information services including navigation, analytics and security.
- Connectivity services – global connectivity anywhere from any device.
- In-space robotics – for science enterprise and consumers.
- Low-cost access to space.

The aim of the strategy is to double the value of space to wider industrial activities from £250 billion to £500 billion, generating an extra £5 billion in exports and attracting £3 billion in inward investment. The goal is to encourage diversity and inclusion in the workforce of the space industry and to interact with 1 million younger people per year to increase the interest in science, engineering, mathematics and technology (GovUK, 2018).

This 'new space revolution' and 'prosperity from space' initiative generated various excited comments from industry and government representatives.

> Too few people are aware of the success and significance of the UK space sector, of its vast potential for future growth. The government welcomes this contribution from the UK space industry and is determined to support our emergence as a commercial space service hub with an increasing global market share backed by new partnerships in science and trade. (Graham Stuart, Minister for Investment, 2018)

> We want the UK to drive the commercial space age and have committed £150 million in our Industrial Strategy to help develop advanced rocket engines, test satellites and establish space ports in the UK for the first time. The government will continue to work closely with the space sector to build on our significant capacity and maximise the benefits of space to life on Earth, creating jobs and opportunities across the country. (Sam Gyimah, The Science Minister, 2018)

In June 2020: 'the UK and US governments signed a new agreement paving the way for US companies to operate from UK spaceports and export space launch technology...Bringing launches to the UK will also be a catalyst for the growth in the UK space industry currently employing 42,000 people and generating an income of £14.8 billion each year'. (UK Space Agency, 2020)

Physical locations

UK spaceports are currently designed to launch small satellites into space and are not primarily to launch people into space. However, the spaceports could contribute greatly to support the way in which Western societies nowadays operate and function. Location-wise, the horizontal launch sites – where the plane takes off from a conventional runway, carrying a rocket with it, and then when the plane is in flight launches the rocket into space to deploy a satellite or space tourism vehicle – will be located in Cornwall, Newquay and Glasgow Prestwick Airport, the last one being the only airport where passenger space flight will also be possible alongside the launching of satellites and microgravity experiments. These horizontal launch sites also follow the principle of sustainability as they can be integrated into existing airport facilities, provide lower costings than vertical launches, can be carried out more frequently with less noise and do not require extensive new developments (Hutton, 2019).

The space hub in Sutherland will be a vertical launch site in the Highlands (A'Mhoine Peninsula) where a rocket is designed to take off upright and will be used for launching small satellites with an expected six launches per year. Making news on the sustainability front, its new Prime Rocket is designed to be used at the site and is forecast to be the first commercial rocket engine to use bio-propane, which has lower carbon emissions compared to traditional rocket fuels (HIE, 2019).

However, some local residents near the future spaceports, such as Space Hub Sutherland in the Scottish Highlands, have already raised concerns over the impact on the environment, local roads, noise levels and crofting rights. The Highlands and Islands Enterprise has therefore needed to provide planning proposals following community engagement and consultation, as well as community newsletters and a FAQ page, to provide more information about proposed plans to locals. Conditions attached to the board's approval included the need to identify and deliver local community benefits as well as focusing on delivering economic benefits, including jobs, training and supply chain opportunities. Detailed gathering of evidence will be vital to ensure that the environmental and community impacts are fully assessed. As the project needs to meet strict regulations, it is subject to a rigorous planning process and must give local people and those involved in taking the project forward a chance to share information, review process and discuss all aspects (HIE, 2019).

Global Spaceports

In addition to the UK, a number of other countries are also involved in developing the space-related industry. In 2017, orbital launch attempts were accomplished by the global leader the United States (with a 32% share), Russia (23%), China (20%), Europe (10%), Japan (8%), India (6%) and New Zealand (1%). The US SpaceX Falcon 9 rocket took over the position of the Russian Soyuz, which had previously flown more times than any other launch system, and led the way to more sustainable operations in the future as it managed to reuse cores that had flown previously (Spaceflight101, 2019).

The United States

In the United States, FAA-licenced spaceports are approved to operate as commercial spaceports. Most of these non-governmental facilities have grown from former military and NASA installations, for the obvious reason that setting up new space-related infrastructure is extremely costly.

- The California Spaceport is the oldest licenced commercial space launch site operator in the United States, operating without government funding and capable of supporting both polar and ballistic trajectories using smaller Minotaur class boosters.
- Mojave Air & Spaceport has become the country's premier test site for private space vehicles and was the launch site of the first commercial space journey, on SpaceShipOne, which won the Ansari XPRIZE in 2004 (Mojave, 2019).
- Oklahoma Spaceport is the first inland spaceport to completely avoid military and restricted airspace and has a commercial facility suited for commercial horizontal take-offs (Osida, 2019).
- Cape Canaveral Spaceport has been operated by NASA, but after the reduction in shuttle programmes in the 2010s, both Cape Canaveral as well as the Kennedy Space Centre have opened up to additional endeavours, together facilitating three active launch pads and two active runways for horizontal launches (Kennedy Space, 2019).
- Spaceport America in New Mexico can be considered the world's first purpose-built commercial spaceport and is the base for private space companies such as SpaceX, Virgin Galactic, EXOS Aerospace and Energetic Pipeline2Space. The site provides 6000 miles of protected airspace to privately engineer, manufacture, test and launch spacecraft. The economic impact of the spaceport was over $20 million in 2016 and the site has an educational programme called STEM Education that aims to inspire and educate the next generation about space exploration and commercialisation (Spaceport America, 2019b).

Russia

The Baikonur Cosmodrome is the world's first and largest operational space launch facility, leased to Russia until 2050 and located in Kazakhstan. It was originally built in the 1950s as the base of operations for the Soviet space programme and has numerous commercial, military and scientific mission launches annually. The first human spaceflight, involving Yuri Gagarin, was launched from Baikonur as were many other historic space flights, such as the launch of the first manufactured satellite and the first space tourist Dennis Tito. Baikonur has also been a major part of Russia's contribution to the International Space Station and, with the conclusion of NASA's Space Shuttle programme in 2011, it became the sole launch site used for crewed missions to the International Space Station (NASA, 2019l).

As the Baikonur spaceport has existed for decades, some environmental research on the impact of the site has been conducted. In the 1990s, research was carried out on the mass deaths of birds and wildlife in the Sakha Republic along the flight paths of rockets launched from the Baikonur Cosmodrome. Scientific studies concluded that the rockets did have effects on the environment and the health of the population, also causing acid rains and increased cancers in the area. Following this discovery, some of the flight paths were changed, but the remains of the polluted scrap metal that had fallen to the ground are still part of the local recycling economy. It has also been confirmed by the Russian Space Agency (Roscosmos) that rocket fuels have been environmentally harmful as heptyl has not been replaced with other substances. However, the site has professed to improve the environmental sustainability of the existing launch vehicles. Such an agreement enabled Russia to continue to 'rent' Baikonur to NASA and ESA as they are still using the launch site for their spacecraft (Viikari, 2007).

Russia's other spaceport, Vostochny Cosmodrome, which is still partly under construction, had its first launch in 2016 and was built to reduce Russia's dependence on Baikonur and to enable Russia to launch most missions from its own land. Apparently, there have been some instances of corruption in the development of Vostochny, but it is expected to have a positive impact on the economy of the relatively poorly developed Russian Far East. Furthermore, several enterprises involved in human space flight are expected to move their activities there after the port is fully completed (Wikipedia, 2019c).

China

Today, Asian competitors such as China and India are entering the space tourism business, changing the market from the previous backdrop of Cold War competition between the United States and the Soviet

Union. By 2045, China aims to become an all-round world-leading country in space equipment and technology and, with advanced space transport capabilities, China will be able to carry out large-scale exploration on planets, asteroids and comets in the solar system (Ma, 2017).

Wenchang Launch Centre on Hainan Island was completed in 2013 to enable China's space station ambitions and boost plans for interplanetary exploration. The facility has been built on a relatively underdeveloped north-eastern corner of the island, but instead of preserving the launch site's isolation, city planners are aiming to integrate the spaceport into the island's tourism infrastructure, including a space-related theme park and 37 other development projects. The local development plans involve the infrastructure to support both the space professionals living and working on the spaceport, and also affluent Chinese wanting to 'take in a space launch while enjoying a holiday on the beach'. The new launch facility symbolises the future of China's space programme not only because of the significantly increased capacity of the new series of wider-bodied rockets that will be launched from there, but also because of the changes in Chinese space culture that the new facility represents (David, 2014).

China has plans to build a spaceplane to accomplish suborbital flights carrying up to 20 tourists to the edge of space. The China Academy of Launch Vehicle Technology (CALT) has designs for a spacecraft like US Space Shuttles, without external fuel tanks, reaching 100 kilometres in altitude, followed by an unpowered descent and glide to an airstrip landing. The spacecraft could give 20 tourists a four-minute weightlessness experience. It would be designed to be reused up to 50 times before decommissioning or refurbishing. This could make suborbital flights affordable or at least within the same range as Virgin Galactic's offerings. China has also announced plans to build a nuclear-powered shuttle capable of deep space flight to asteroids and sees a nuclear fleet as the key to interplanetary flight, commercial exploitation of near-Earth and to colonise the solar system (21stcentech, 2019; Wikipedia, 2019d).

India

India's Department of Space has a vision 'to harness space technology for national development, while pursuing space science research and planetary exploration' (ISRO, 2019). Satish Dhawan Space Centre is a rocket launch centre operated by the Indian Space Research Organisation and the first orbital satellite was launched there in 1979. It is the main base for the Indian Human Space Flight Programme and the existing launch facilities are being augmented to meet the target of launching a crewed spacecraft called Gaganyaan in 2021. If completed on schedule, India will become the fourth nation to conduct independent human space flight after Russia, the United States and China, aiming to start a space station programme and possibly a crewed lunar landing. Similarly to China, India could offer tourism opportunities through its space programme, including suborbital and orbital travel. However, India is yet to launch a human into suborbital or orbital space using a rocket of its own design so suborbital space tourism flights may still be 15 years away (ISRO, 2019).

Japan

The Tanegashima Space Center is the largest rocket launch complex in Japan, located on an island on the southeast coast of Tanegashima. So far, the spaceport has been used mainly for testing, launching and tracking satellites, and also includes on-site facilities such as The Space Science and Technology Museum (Wikipedia, 2019e). The Japanese Aerospace Exploration Agency (JAXA) has, however, set a target year of 2022 to launch astronauts aboard two possible technologies, a Japanese-made capsule and a space plane, designed to carry a crew of three and 400 kilograms of cargo. In 2018, Japan's HTV-7 cargo delivered resupply capsules to the International Space Station to test the capsules' recoverability and usability. Japan also has ambitions to start offering a point-to-point suborbital passenger service with new spaceplane technology within a decade (21stcentech, 2019).

Conclusion

The human era, the Anthropocene, has altered the world beyond anything in its previous geological timescale. The industrialisation of Western societies has impacted upon our planet, with climate change being one of the most prominent consequences. There have also been changes in society reflecting on the tourism industry as the paradigms have moved towards tourism degrowth and global equality. In order to survive, businesses of the future need to form their operations and relations in line with global environmental megatrends and comparative fairness – especially if wanting to attract the younger generations.

The emergence of the space tourism industry will have various impacts on local areas: in positive terms it will provide employment opportunities but, if not planned properly, it may possibly have a negative impact caused by the formation of a new tourism destination. Therefore, it needs to be stressed that social sustainability factors should be strictly regulated by local and national authorities and it would be advisable for authorities to emphasise the moral and ethical code of conduct, in order that international businesses, such as space tourism companies, succeed in the long term in a local area. Building and creating all the necessary infrastructure for spaceports require a vast amount of economic capital and there is lots of bureaucracy involved, so for space tourism companies it would be a very far-reaching and costly mistake not to guarantee the support of the local community.

6 Economics and Space Legislation

> The space tourism venture is essentially a bet on the fast-growing luxury experiences market. Globally we think around 2 million people can experience this over the coming years at this price point.
> George Whitesides, CEO Virgin Galactic, 2019

Introduction

Expenditure on the unique experience of space tourism promises to play a positive role in the economy and society, enriching customers culturally without requiring mass production of consumer goods and the corresponding pollution, thereby acting as a harbinger of a future open world economy (Collins, 2003). The future earning opportunities in the space industry will be regarded as highly desirable for many companies, such as space tourism operators, especially if the practicalities can remain sustainable or be developed sustainably to match the global environmental consumer megatrends. However, many argue that the entire idea of sustainability is incompatible with that of economic growth (Viikari, 2007). As space is becoming increasingly commercialised, new global legislation is needed to regulate operations, particularly those of the new space economy participants, to legally bind space practices to sustainable criteria.

This chapter explores the new space economy and the impact of reusability on the commercial space sector. This is followed by a study of the space tourism economic forecast, pricing and the phases of the cost structure development. Insight into the future space tourism market and its demand is then analysed and finally, the existing space legislation is explored, with attention given to the elements of sustainability, ethics and safety.

New Space Economy and Space Reusability

One of the principles of sustainable development is to balance environmental protection and economic development in a way that is sustainable for the future of humankind. The utilisation of space tourism's potential for the development and operation of next-generation launchers was brought into focus in the 1980s. There was an emerging discussion as to whether it would be feasible to establish a positive interaction between the obvious demand for space tours and the need for vast markets to achieve reusable launch vehicle operation cost targets. The era of the commercialisation of space has brought new, more economically advanced alternatives to previous governmental operations, referred to as 'new space', which will have substantial effects both within and beyond our biosphere (Spector & Higham, 2019). An existing example of sustainability in new space is satellite monitoring of climate-related environmental change. However, ironically, the process of escaping the confines of the biosphere has thus helped reinforce the notion of planetary boundaries and limits (Spector *et al.*, 2017).

Since 2010, the space tourism industry has been moving towards cheaper access to space by using reusable launch vehicles. For example, SpaceX and the Indian Space Research Organisation have been successful in a series of technology demonstration missions realising reusable launch vehicles. Although such sustainable design requires a very high initial development cost investment and needs investors, it is the only option promising a substantial reduction in cost per launch. This will also bolster economic success, increase commercial business within the aerospace industry with its widely discussed positive implications and satisfy the customer megatrend for more sustainably operated vehicles. However, it needs to be noted that the reusable design can only accomplish its low-cost target if it can reach high launch-per-year rates (Abitzsch, 1996; ISRO, 2019; SpaceX, 2019b).

It is already apparent that there are enormous price differences between different private space industry providers. For instance, it has been calculated that sending an astronaut into space with SpaceX Crew Dragon costs $55 million whereas Boeing Starline will charge $90 million. The testing processes of the new space fleet cost $2.2 billion for the Boeing Starline and $1.2 billion for the SpaceX Crew Dragon (Keränen, 2019b). The biggest reason for such a marked difference is that SpaceX uses reusable launchers whereas Boeing's space fleet operates within a less economically sustainable framework. Thus, it can be concluded that incorporating sustainability into the operational business model of commercial space tourism provides advantages, such as more competitive pricing, to the end user. However, as these vast figures indicate, space tourism already suffers from the real danger of extreme manifestation of the global trend that is essentially a feudal system, with a very elite few being able to go to space due to the efforts of the masses.

In 2017, the number of commercial launches started to approach that of those conducted by government organisations. For example, SpaceX successfully completed 18 Falcon 9 missions while the United States conducted 29 orbital space launches using different launch sites and operators. SpaceX launches included a variety of products, ranging from orbital logistics, high-energy geotransfer missions for telecommunication operators and launches for the US government to flights supporting the observation of Earth. Of particular note is that SpaceX managed to reuse the space rockets, completing 14 successful first-stage recoveries at a 100% success rate and flying five missions with previously used boosters to take its reuse ambitions a step closer to becoming routine. Progress was also made in fairing recovery endeavours with expectations of achieving intact fairing returns on a regular basis in the future (Spaceflight101, 2019; SpaceX, 2019a).

Commercial space travel and its numerous spin-off activities have the potential to escape the limitations of consumerism, which governments in rich countries have encouraged in recent decades in order to stimulate economic growth. This has resulted in excess consumption, causing unnecessary environmental damage, while reducing, rather than increasing, popular satisfaction (Cole, 2015). However, the environmental effects of space activities often involve a time frame that will be considerably longer than in most other investment decisions. It is very likely that it will be future generations that obtain the benefits, rather than the generation that incurs the potentially very high starting costs.

In the future, numerous private stakeholders will be primarily concerned with getting their share of the benefits of space activities, possibly creating an issue around 'free-riding' when it comes to environmentally sustainable activities, meaning that many will assume that others will make enough effort to increase environmental safety and protection, from which they too can benefit, without shouldering any extra costs (Viikari, 2007). In the long term, reusable space vehicles will be able to offer cheaper access to space, boosting the space tourism market itself and, as a result, giving rise to a sustainable crewed space transportation market. The perfect symbiosis could be established between the next generation of low-cost launch vehicles and sustainable space tourism.

Economic Forecast

As with any rapidly advancing technology, such as the developments in computer and mobile phone products, there are likely to be price reductions as the technology matures, and the scale of operations, operators and markets increase. Over the last decade, the precise costs of space tourism have varied, but the reason for this is not yet clear; it may be due to modified vehicle techniques and spaceport locations, discount rates and margins or perhaps to products and market appraisal (Cole, 2015).

It is a complex undertaking to evaluate the economic revenue forecasts for space tourism, as at the time of writing (March 2020), the commercial space tourism industry has not yet started to operate on a so-called 'elite mass' scale, and instead is accessible only to astronaut tourists such as Dennis Tito. Forecasts predict that space tourism will become a very lucrative private business, with an annual market of several billion US dollars and significant scope for expansion (Global Information, 2018). In 2017, almost $4 billion of private money was invested in space projects, almost entirely in the United States (Tett, 2018).

UBS (2019) financial services in the United States estimates that space tourism will be a $3 billion market by 2030. Private space companies are currently investing heavily in space opportunities and space is seen as an enabler for broader investment opportunities. Another spin-off enhanced by space tourism, superfast travel using outer space, could be a $20 billion market and compete with long-distance airline flights. Point-to-point flights with rockets could make 10-hour plus long-haul airplane flights practically non-existent. The broader space industry, worth about $400 billion today, has been estimated to double to $805 billion by 2030 with these space travel innovations included (UBS, 2019). The UBS analysts justify the figure estimates as follows: 'while space tourism is still at a nascent phase, we think that as technology becomes proven, space tourism becomes more mainstream. Space tourism could be the stepping stone for the development of long-haul travel on Earth serviced by space' (Sheetz, 2019b).

Richard Branson's Virgin Galactic Holdings became the first space tourism business to go public in October 2019 (Saigol, 2019). Analysts believe that Virgin Galactic represents an opportunity to invest in three areas: the rapidly growing luxury consumer market, the pioneering of new technologies and the popular theme of experiences over possessions (Sheetz, 2019a). The safety factor of the flights is estimated to have a particular role in Virgin Galactic's stock price as the market appears to imply a high probability of failure and investors are likely to use a crash rate similar to that of the Space Shuttle (two fatal accidents, equating to 13% of journeys) to estimate Virgin Galactic's potential failure rate. After SpaceX's Mk1 booster rocket exploded on the testing site in Texas in November 2019, the company's CEO Elon Musk quickly announced that SpaceX will no longer continue to use this rocket (Tuohinen, 2019).

BOX 6.1 ANALYSIS OF VIRGIN GALACTIC SHARES

- Virgin Galactic became the world's first and only publicly traded commercial human space flight company on the New York Stock Exchange on 20 October 2019.
- The share price soared over four months from $11.75 to $37.35 (20 February 2020).
- Revenue is expected to grow from $31 million in 2020 to $590 million in 2023 as flights grow annually from 16 to 270.

- Virgin Galactic's business is based on two companies: the consumer-facing flight company Virgin Galactic and the aircraft design and manufacturing company The Spaceship Company.
- Since the manufacturing company is developing its own technology, it can sell aircraft to third parties, develop high-altitude platforms and progress into electric air mobility.
- As the ticket price is $250,000, the addressable market is very small.
- There are 1.78 million people with a net worth of over $10 million, which is the first market.
- Virgin Galactic already has $80 million in deposits so there is clearly a demand.
- These early customers provide the market, pay for upgraded aircraft and give the company scale to grow the fleet to an estimated five spacecraft by 2023.
- Over time, ticket costs should come down as efficiencies and scale help the business and the potential market expands.
- There are 37 million people with a net worth ranging from $1 million to $5 million, which could be the second market.

(Source: NASDAQ, 2020)

Space Tourism Pricing

As early as 1996, questions were raised about the prospect of space tourism in the European Aerospace Congress, including:

(1) When thinking of the price development of future space tourism, will the cost of offered space trips fit into the demand price figures?
(2) Will there be enough affluent customers willing to pay high-price tickets until the price comes down to more affordable levels for the majority?

A sustainable space tourism market will emerge when the demand price lies above the supply cost. Otherwise, this 'price gap' must be bridged or the general public will have to accept that a self-sustained market will never realise. If the first ticket prices are forecast to cost about $200,000, with a volume of 1,000 passengers per year and the price then drops to $100,000, it may be possible to establish a volume of 10,000 to 100,000 passengers per year. It is interesting to note that in this price regime, the US market alone will account for up to 30,000 space tourists. The maximum market potential considering these figures will emerge at ticket prices of approximately $35,000 with 2,000,000 space tourists per year. That would produce an annual turnover of $70 billion (Abitzsch, 1996). Alternatively, a possible 'chicken and egg' problem could also be created: (1) high launch costs could lead to low transportation demand and (2) low demand could result in small markets with low turnover – such a scenario would not convince any aerospace industry or private investor to invest in expensive new launcher developments, resulting in the access-to-space costs remaining high and markets small in the future (Abitzsch, 1996).

Up to 2019, less than 10 private individuals have paid the fee of around $20 million to travel to space and to the International Space Station. This figure will start increasing when NASA begins its commercial tourism to the International Space Station in 2020 and Virgin Galactic, which has already accepted more than $80 million in deposits, starts to operate suborbital space tourism. Currently, Virgin Galactic (2019b) has 603 customers signed up to fly once it begins commercial operations next year, at a price of $250,000 per ticket. NASA will allow private citizens to visit space for trips up to a month-long to the International Space Station, with an estimated cost of $50 million per seat. In addition, NASA will charge visitors for basic services such as food, storage and communication at the station. However, such journeys are planned to take place only twice a year so the activity will have a limited profit margin (Mansi, 2019).

SpaceX's CEO Elon Musk has estimated that the price of a seat aboard SpaceX's Starship interplanetary vehicle will eventually drop enough to be available for mass tourism from industrialised countries.

> Very dependent on volume, but I am confident moving to Mars (return ticket is free) will one day cost less than $500,000, even below $100,000. Low enough that most people in advanced economies could sell their home on Earth and move to Mars if they want. (Elon Musk, 2019, in Wall, 2019b)

Blue Origin's pricing policy is so far forecast to be close to that of Virgin Galactic's suborbital flights, costing about $200,000–$300,000 per ticket. The tickets are not yet available for purchase and it has been estimated that, initially, the company is likely to lose as much as $10 million per flight as the New Shepard system can only carry up to six passengers 62 miles above the planet's surface. The space tourism capsules are designed to have 'the biggest windows in space' to attract space tourism enthusiasts from all over the world (Moon, 2018).

BOX 6.2 COST STRUCTURE PHASES

(1) The first phase will be an 'exclusive' one, targeting the wealthy elite and offering short space jumps. It has been estimated that even within five years of such operations, a cost decrease will allow

substantial price cuts, meaning that there would be a larger potential customer base able to afford a space trip.

(2) In the next 'executive' phase, following perhaps 5 to 10 years later,
the space tourism market would witness even lower prices with a
relatively high turnover, as businesses begin to target more towards
the mass market. This phase would be a potential time to implement tourism infrastructure, such as space hotels, to ensure both
the spurring of new demand from the exclusive market segment,
and later on enhance the base of customers.

(3) Supporting activities to stimulate demand could then be implemented and the most promising could be a space tour lottery for
social media influencers, providing an opportunity to experience
space travel if otherwise financially not possible. These people
could fly earlier than expected and, by sharing their experiences
with millions of worldwide social media followers, could boost
demand and hence support a price decrease owing to the economy
of scale.

(4) In connection with the development of an advanced lunar base,
passenger flights to the Moon or even Mars could be offered at
relatively low cost, with a passenger round trip from the Earth to
the Moon costing less than $500,000.

(Source: Abitzsch, 1996; Collins, 2003; Collins &
Autino, 2010; SpaceX, 2019a)

Space Tourism Market

Plog's (1974) heuristic vision, describing how the distribution of
visitors' personalities along a psychometric spectrum, from very cautious
and conventional to risk-taking and adventurous, overlays the positioning of destination, could be used as one tool to observe the different
development phases of space tourism. For example, in the future, the
same personality types now conquering Mount Everest would be similar
to the first-phase pioneers to set foot on the Moon or Mars. In the first
pioneering phase, space tourism will be a luxurious and exclusive, one-
off thrilling experience for the world's wealthiest elite. However, it will
still compete with other adventure tourism activities such as mountain
climbing and deep-sea diving, considered to offer better safety features
and value for money.

Ormond and Dickens (2019: 233) define three types of personalities
in space pilgrimage:

• Those whose motive is to leave the familiar structure of the Earth's
society and social relations in order to remake the self. For these pil-
grims, space tourism represents a life-transforming experience.

- Those wanting to experience the thrill of a once-in-a-lifetime adventurous experience, reinvigorating their life with a sense of awe and mystery not possible in everyday life.
- Those seeking public respect and admiration for their bravery to accomplish such a dangerous journey.

Space tourists of the future may also travel to space simply to demonstrate that they are able to afford the latest adventure tourism extravagance (Ormond & Dickens, 2019).

As technology keeps advancing, the potentially very large future market will eventually start reducing the cost of space travel in a manner similar to that of large-scale operations like airlines, which have already demonstrated that once the technologies reach maturity, the costs fall drastically. For example, a cost comparison indicates that the price of a seat on a Boeing 747 is about 1000 times less than that of a seat on an orbital journey on the Space Shuttle (Cole, 2015). However, it is important to remember that as the cost of space tourism becomes less expensive, the experience will be extended in both distance and duration and trips may become more costly. There could, for example, be substantial cost reductions in moving from small space vehicles to those with larger capacities, and the main costs would be for operating the vehicles rather than increasing the price for space tourists. There could also be some advantage to sustainable infrastructure in joint launch sites for smaller-sized spacecraft, suggesting that several independent commercial enterprises should coexist, at least in the pioneering phase of space tourism.

Market Demand

In theory, suborbital space travel services could have started as early as the 1970s, as the basic technology required for space trips was already available. From an economic point of view there has already been a large cost to not benefiting from the commercial space activities that could have been generated so far. Collins (2003) speculates that $1 trillion of public expenditure could have come directly from private industries, were it not for the delay in the development of the commercial space industry, and that the resulting commercial turnover is about 1/50 of what ordinary companies would have achieved with the same investment. Also, another $20 billion per year is spent on non-science space activities with little economic value, and this, in addition to the economic, social and macro-economic costs, has created lost potential value of new economic activities.

One could speculate that there is a great unsatisfied demand for the idea of space travel, as the topic of space has, for decades, raised interest among the public due to Hollywood productions and other interpretations. As the commercial space tourism industry has not yet started to operate, there has been very limited research, simply because space tourism has not, thus far, been an option, making it difficult to survey the public. So far, for the average person, one has needed to train to become an astronaut in order to have a chance to experience a space trip.

Over the decades, some surveys have estimated the potential demand for space tourism. For example, in 1993, Tokyo University with P. Collins conducted research titled 'The Evaluation of Global Space Tourism Demand between Japan, Germany and USA' with the following key findings:

- Among those three nations, the most enthusiastic future space tourists appeared to be the Japanese (70%) whereas the Germans showed the lowest percentage (43%); however, all nationalities were quite supportive of the prospect of space tourism.
- The preferred activity during a space trip was dominated by one pursuit – looking at Earth – while other activities that could be realised without great expense included astronomical observation and a G-force experiment. Spacewalking was also a popular choice.
- It was, however, noted that space tourism does not necessarily need to include expensive high-tech attractions that consume additional payload capacity. Ensuring a good visual of the surroundings with an adequate number of windows and an attractive flight profile (e.g. highly inclined orbit) to see as much as possible of the Earth was highly recommended.
- It was also noted that despite cultural differences between Asian, European and American people, the similarity of the survey results was striking. This would especially favour space tourism entrepreneurs, because no specific offerings need be planned according to the origin of passengers.
- The preferred length of space trip was several days for most of the respondents, suggesting that there would be a need for orbital accommodation enabling people to stay a few days in space and that without such infrastructure the demand for space tourism would not reach its full potential. This, however, will be challenging in the initial phase of space tourism as the first years will be dominated by short trips into low-Earth orbit.
- The sum that people were willing to pay for a space trip was quite similar among all nations, and equated to one month's salary; however, 5% of the Japanese respondents were even willing to pay five years' worth of their salary for a space experience.
- On the other hand, the main reasons that people were not interested at all in space tourism included safety concerns from a third of all respondents, regardless of nationality, with a similar percentage preferring simply to dream about the experience.
- Respondents reported that the stage at which space tourism could become a real option would be when safety concerns are overruled by

confidence-building safety measurements and also when space tourism simply becomes a reality.
- Nearly 10% of the respondents, despite the country of origin, did not find the idea of space tourism interesting, as it was not considered a realistic travel option.
- The estimated global market potential was calculated as passengers per year versus the ticket price and the result was that the maximum potential lies at 20 million passengers per year; however, this requires a ticket price of $1000 or less and is not likely to be achieved anytime soon.
- A price of more than $100,000 would not lead to a sustainable market as the demand would fall to a low number of passengers. A self-sustaining space market could be expected with ticket prices of $50,000 or less taken into account as the primary factor for the future space vehicle design process.

In 2002, a comprehensive space tourism market study was accomplished by Futron Zogby. It presented the demand for space travel from 2002 with a 20-year forecast. A telephone survey of 450 households in the United States with a yearly household income from $250,000 focused on defining three key aspects limiting tourist demand: fitness, training time and cost (Beard & Starzyk, 2002). The findings revealed that:

- 19% were willing to try a suborbital flight.
- 42% said seeing Earth from space was the main reason to travel.
- 35% were motivated by being a pioneer in a new activity.
- 32% were willing to pay only between $5,000 and $10,000 for a suborbital flight.
- 14% were willing to pay more than $50,000 for a suborbital flight.
- 67% were 'definitely likely' to spend months in training for an orbital flight, compared with only 22% for a suborbital flight.

In 2019, some general attitudes towards space tourism were measured in a web-based survey in Finland (Toivonen, 2019), with a total of 132 respondents representing an age range from 18 to 75. They were asked to provide their views on different questions on a Likert scale, with options of 'strongly disagree', 'disagree', 'neutral', 'neither agree nor disagree', 'agree' and 'strongly agree'. Some of the most popular views are summarised below. Other responses were quite evenly divided across the other Likert options, with the exception of questions related to the environment and legislation.

- 32% agreed that the idea of space tourism is interesting.
- 27% strongly agreed that they would participate in a space trip if it was economically possible.
- 29% remained neutral when considering whether they would like to try space tourism even though they would not consider it safe.

- 40% remained neutral on the prospect of joining a space trip only if it would not pollute the environment or they could participate without physically travelling themselves (such as virtually).
- 44% remained neutral on the idea that the start of space tourism in the next few years represents good progress for humanity.
- 43% agreed that they were worried about the emissions caused by space tourism.
- 47% strongly agreed that space tourism should be regulated by international agreements.
- 52% strongly agreed that humankind does not have the right to litter the space environment.

The findings indicated that the general public in industrialised Western countries may still find the option of commercial space tourism quite distant from their daily lives, even though as an idea it raises common interest. The megatrends of sustainability and concern about climate change were quite visible, as half of the respondents agreed strongly on the topic of human responsibility while utilising the space environment, and also supported stricter international legislative regulations. Once the commercial space tourism industry starts operating, it will be essential to carry out more research also based on the opinions of pioneer space tourists and to include research on a range of related matters, such as sustainability issues, passenger space flight services and even orbital accommodation as desired from the point of view of space passengers.

BOX 6.3 MEETING THE DEMANDS OF THE SPACE TOURISM CUSTOMER

- It is essential to meet the expectations of space trip participants, by understanding, for instance, that most tourists want to look at the Earth and experience weightlessness. Vehicle design should therefore provide a sufficient number of windows and interior space in which to float around.
- High inclination orbits should be avoided as they cover a greater proportion of the Earth's surface.
- The acceleration level should be kept lower than 3G (due to medical restrictions).
- The first space trips should last a maximum of a couple of hours to avoid space sickness.
- The space flight time should, however, be long enough to create a feeling of 'value for money'. If not, to compensate for a shorter flight time, a luxurious or adventurous space camp could be implemented before and after each space flight. This could intensify the feeling of becoming a 'real astronaut' as the camp would provide

some training and medical check-ups as well as a sense of customer satisfaction.
- All appropriate procedures for health and safety are essential. Final health inspections should be completed just before a space flight to avoid any unnecessary incidents.
- Continuous monitoring of customer demand and pricing is essential to ensure the continuity of the space tourism market.

(Source: Adapted from Cole, 2015)

Space Legislation

> The future space legislation should expand the American experiment and environment that led to not only pioneer railroads, automobiles, and planes, but to turn these technological advancements into sustainable businesses. (Eric Stallmer, President of the Commercial Spaceflight Federation, 2015)

Over the decades, space exploration has created new understanding and scientific data on the Earth and remote regions of the universe and given rise to new trends in economic activity, which today are unthinkable without the use of space technology and communications, meteorology, navigation, satellites and the study of the Earth's natural resources. Many new technological processes are likely to evolve into space-based branches of production. Modern space activities have also included large and multinational businesses, which typically have tended to result in accelerated environmental destruction as the ideology in private space businesses revolves around the maximisation of economic profit (Viikari, 2007).

From the very beginning, space law has been called upon to secure the stable and peaceful development of both national and international space activity and to meet the interests of both individual countries and the world community as a whole (Vereshchetin *et al.*, 1987). The legislation and jurisdiction as they exist on Earth require a different approach in space as international laws have not yet been similarly developed. Even the process of physically determining where space actually starts has brought political and legal issues, because aerospace partly falls under national sovereignty whereas outer space does not (Viikari, 2007). The application of air law and space law still needs to be agreed upon.

Space tourism represents an idealised experiment for international and domestic policy implementation as it in one part determines whether private and public valuation of an environment can coexist. It should be regulated as a subset of private space flight activities, although both the existing space law and air laws have yet to have international-level regulated activities to any appreciable extent. There are current discussions about which laws and regulations should apply to private suborbital space flight and space tourism, and which should be further developed to fit the future variety of space flights in any existing category, either within space activities and space law or within aviation and air law. According to von der Dunk (2019: 182), 'the only thing which can be safely said at the outset is that the global character of both space activities, and generally, aviation means that any analysis of existing law and development of future regulation should preferably and primarily focus on international law as opposed to domestic law, limited to respective single nations. In fact, international space and air law serve to determine the scope for such national laws'.

Alongside the development of the space tourism industry, there are also plans for private space mining, which will fall under another legal sector as air law is completely irrelevant to the ownership of space resources (von der Dunk, 2019). The crucial question, however, is, as private space tourism forms itself, how should this growing industry be regulated, especially in terms of sustainability, to harness private businesses and countries globally? Also, will there be a political will and a private sector imperative to mitigate damage to the space environment, and will there be an actual willingness to leave some potential space tourism locations untouched? (Duval & Hall, 2015: 451)

Even though the development of space technology has been fast and impressive, there has not been equal success in learning lessons from terrestrial history regarding the importance of environmental protection, resulting in an increase in space-related environmental problems. The growth in the amount of space debris has created more awareness of the seriousness of the problem, though, and both the governmental sector and the industry have made efforts to mitigate further hazards by developing procedures and standards for the operation and design of space missions (Viikari, 2007).

The basic principles of international law are universal and form the basis for the legal regulation of relations between countries. They are usually applicable to all types of activity by the subjects of international law in the sphere of international relations. However, the general principles, even those applicable to all types of activity, are not enough to govern the exploration and use of outer space since many Earth-bound principles do not take account of the specific features of space activity and thus cannot cover all concrete problems arising from space-related activities (Vereshchetin et al., 1987). As satellite activity comprises a significant portion of current activities in space, and other types of space endeavours such as space tourism have not yet occurred, many of the recently added treaties concentrate on the environmental effects of the use of satellites.

The Outer Space Treaty (1967) corresponds to maritime laws and constitutes the only legislation in existence for common space responsibilities. It includes the principles governing the activities of states in the exploration and use of outer space, including the Moon and other celestial bodies. The treaty was based on the Declaration of Legal Principles

Governing the Activities of States in the Exploration and Use of Outer Space (1963) and entered into force in October 1967 by the governments of the Russian Federation, the United Kingdom and the United States (UN, 2019d).

The body of international space law consists of five United Nations treaties:

(1) 1967 Treaty on Principles Governing the Activities of States in the Exploration and Use of Outer Space, including the Moon and other Celestial Bodies
(2) 1968 Agreement on the Rescue of Astronauts, the Return of Astronauts and the Return of Objects Launched into Outer Space
(3) 1972 Convention on International Liability for Damage Caused by Space Objects
(4) 1975 Convention on Registration of Objects Launched into Outer Space
(5) 1979 Agreement Governing the Activities of States on the Moon and Other Celestial Bodies

(UN, 1979, 2019d)

The space sector has needed to cope with the global differences in development, especially as outer space as an environment and a resource has typically been perceived as some sort of limited pie of rights to which all states aspire. The more-developed nations have been eager to perceive outer space and its resources as common property, available on the basis of the 'first come first served' attitude, whereas the less-developed nations have been concerned about being guaranteed adequate possibilities for equal benefits either in the present or in the future (Viikari, 2007).

Before the Outer Space Treaty was created, there were global concerns that the major space powers, the United States and the Soviet Union, would overshadow and influence the development of space law. Developing countries from Asia, Latin America and Africa in particular were concerned that the benefits to be found in space would remain limited to only a small number of advanced industrialised countries, wanting to also benefit from the new technology for their economic and social advancement. Included in space law is that there should be benefits for developing countries as well, with one title of the legislation declaring, 'For the Benefit and in the Interest of All States, taking into Particular Account the Needs of Developing Countries', as a reminder to the space powers to fulfil their obligation to conduct space activities for the benefit of all countries. However, in the late 1970s, one could already see a change in the economic philosophy, particularly evident in negotiating the Moon Agreement, where the most advanced technological countries saw little merit in accepting new legal global obligations and preferred to cast their international relations in bilateral form (von der Dunk & Tronchetti, 2015).

BOX 6.4 THE OUTER SPACE TREATY FRAMEWORK

The Outer Space Treaty provides the basic framework of international space law, including the following principles:

- the exploration and use of outer space shall be carried out for the benefit and in the interests of all countries and shall be the province of all mankind;
- outer space shall be free for exploration and use by all states;
- outer space is not subject to national appropriation by claim of sovereignty, by means of use or occupation or by any other means;
- states shall not place nuclear weapons or other weapons of mass destruction in orbit or on celestial bodies or station them in outer space in any other manner;
- the Moon and other celestial bodies shall be used exclusively for peaceful purposes;
- astronauts shall be regarded as the envoys of mankind;
- states shall be responsible for national space activities whether carried out by governmental or non-governmental entities;
- states shall be liable for damage caused by their space objects; and
- states shall avoid harmful contamination of space and celestial bodies.

(UN, 2019d)

The UN space treaties make very limited reference to environmental issues. However, in the light of current debate on climate change, this topic should become one of the highest ranking on the agenda to agree on new legally binding international space legislation in the future. So far, the Moon Treaty is the most advanced of the space treaties in an environmental sense, though it only covers general-level environmental protection.

> In exploring and using the Moon, States Parties shall take measures to prevent the disruption of the existing balance of its environment, whether by introducing adverse changes in that environment, by its harmful contamination through the introduction of extra-environmental matter or otherwise. States Parties shall also take measures to avoid harmfully affecting the environment of the Earth through the introduction of extra-terrestrial matter or otherwise. (UN, The Moon Treaty, Art.7.1, 1979)

Beyond this, space is still equivalent to the former 'Wild West', with a 'first come first served' attitude and approach to rights and actions.

The United Nations has been acting as the centre for drafting the norms and principles of international space law and set up a Committee on the Peaceful Uses of Outer Space, composed of over 50 states. The international law includes numerous treaties and agreements aimed at establishing law and order in space activity, particularly trying to minimise the common threat of space being turned into another sphere of military confrontation. The industrial production of space activities has tended to be military induced and nearly all space technology has had the potential for dual-use between the military and civilians (Viikari, 2007). However, ongoing space exploration and the increase in the use of private space enterprise to assist daily life on Earth have created new legislative problems requiring international political and legal settlement (Vereshchetin *et al.*, 1987).

New regulations

The opening up of space for many activities such as space tourism, instead of the originally intended political and military purposes, has made it necessary to specify new types of regulations and legislative frameworks that have not been covered by the space treaty. Some of the newer norms include bilateral treaties regulating space cooperation between states and governmental space agencies (Von der Dunk & Tronchetti, 2015). Since the millennium, the United States alone has formed over 1000 technical and scientific agreements with over 100 countries and organisations, and in 2017 President Trump signed a NASA authorisation bill announcing that the White House will re-establish a National Space Council and continue to support the developing commercial space sector (Cofield, 2017).

> With this legislation, we support NASA's scientists, engineers, astronauts and their pursuit of discoveries. This bill will make sure NASA's most important and effective programs are sustained. It orders NASA to continue transition activities to the commercial sector where we have seen great progress. It continues support for the commercial crew program, which will carry American astronauts into space from American soil once again. It's been a long time. (US President Donald Trump, 2017)

Another example of national-level space legislation is the UK Space Industry Act (2018) which was created to regulate space activities, suborbital activities and associated activities carried out in the United Kingdom. The Act created a licensing framework for space flight activities and sets some conditions that must be considered, including environmental and safety assessments. For example, all spaceports will need to comply with the existing environmental, planning and health and safety legislation (Hutton, 2019).

> **BOX 6.5 UK SPACE INDUSTRY ACT (2018)**
>
> The Act covers a number of areas:
>
> - Regulation of space flight
> - Range control
> - Licences
> - Exercise of regulatory functions by bodies other than the Secretary of State
> - Individuals taking part in space flight activities
> - Safety regulations
> - Security
> - Enforcement
> - Liabilities, indemnities and insurance
> - Powers in relation to land
> - Powers in relation to land supplementary
> - Appeals
> - Miscellaneous
>
> (Source: Legislation UK, 2018)

Ethical considerations

International cooperation in the exploration and use of outer space can be described as joint scientific, technical, economic, political and legal activities of countries and private companies. Future space legislation, which will involve space tourism, should ensure the achievements of space science and technology for peaceful purposes and for the benefit of all countries and peoples (Vereshchetin *et al.*, 1987). There are also active non-governmental and non-industry interest groups, such as The Planetary Society that promotes exploration of the solar system and the search for extraterrestrial intelligence, which consider themselves stakeholders in the space sector and thus aspire to influence policymaking in space-related matters (Viikari, 2007).

Nowadays, many countries and private companies launch various space objects and vehicles to conduct space research, organise international crewed flights or use space for new commercial services via a satellite network. For example, in November 2019, SpaceX launched 60 new broadband satellites to create the Starlink network, the world's largest commercial satellite network. The satellites are positioned to stay at an altitude of only 280 kilometres, as a lower orbit accelerates the removal of faulty satellites from orbit due to atmospheric friction (Keränen, 2019a).

However, there is still limited international law to regulate similar space actions and sustainable decision-making is largely the ethical responsibility of the private company. In September 2019, SpaceX missed a satellite collision warning because of a fault in software not alerting a human operator, and a European Earth satellite needed to perform an evasive manoeuvre to make sure it did not collide with it. Such manoeuvres may become common in the future as private internet megaconstellations start shaping (Wall, 2019a). If a collision were to occur, it could create a lot of space debris, which may complicate and even endanger the safety of other space-related activities such as space tourism.

To ensure that space will not be commercialised for the powerful and the rich to the exclusion of others and that the commercialisation of space does not breed a new generation of pirates, action needs to be taken by both private and international organisations to regulate space businesses, considering especially the legal challenges to what are thought to be transgressions by businesses operating in space, and governments controlling access to space. According to Marsh (2006), there are already three proposed achievable guidelines for space commercialisation:

(1) *Space preservation*: Space is valued for its own sake, regardless of any benefits that may be derived from it.
(2) *Space conservation*: Protecting and caring for the universe's resources for the sake of others and avoiding exploiting it to benefit only a few.
(3) *Space stewardship*: Holding ourselves accountable for managing space resources. This approach would require that we consider how our actions affect others, our environment and the future.

Safety considerations

The safety of crewed flights in outer space has both national and international aspects. The internal jurisdiction of the state of registration covers the procedure for preparing for flight, even if nationals of other countries are taking part, and all relevant questions are governed by national laws and other legal enactments. When necessary, the country of registration may determine with other countries' regulations their relations connected with flights in which nationals of other countries participate. The international aspect of law is very important for the global nature of space activity as it places in the forefront the need to create favourable conditions and maximum safety for operations (Vereshchetin *et al.*, 1987).

When vehicle movements in outer space start to cause congestion, there should be legal adaptation of certain traffic rules for space routes in order to ensure the maximum safety and efficiency of space tourism flights. Ensuring the safety of space crews and passengers presents one of the key legal problems of space tourism flights as, once a person is launched into space, there is a risk of numerous unexpected situations due to the substantial difference between space activities and any other activity. The safety of crewed space flights should facilitate the prevention of pollution and contamination of outer space, ensure functioning com-munications with all space vehicles and prohibit 'hostile' interference between different spacecrafts from various sources.

There would be a need for an international space rescue service, an international body to regulate questions of safety for space flights or an international space agency vested with coordination and opera-tive functions. This body could assist at outer space accidents involving either astronauts or tourists, in the event of emergency situations such as illnesses or equipment failure, and quickly approach a space object in distress, dock with it and render assistance. The creation of such an inter-national rescue service will, however, require settlement of a number of political, legal, economic, technical and scientific questions (Vereshchetin *et al.*, 1987).

Conclusion

The United Nations Environmental Programme extended the sus-tainability concept in Bruntland's commission, 'Our Common Future', to intra-generational and economic equity by stating that sustainable development requires the use of natural resources in a way that underpins ecological resilience and economic growth, as well as helping to achieve international equity (Birnie & Boyle, 1992). On the eve of the new space tourism industry, responsibility for the consequences of the growth of new space economy businesses, and what this means for different societ-ies on Earth must be carefully considered.

The space tourism industry has the genuine potential to generate the large-scale launch activity needed to reduce costs sufficiently to allow the use of space resources, making it one of the most important projects in the world today (Cole, 2015). The future industry will be part of the new space economy, with enormous economic potential, and the aim to not just create an expensive 'selfie-in-space moment' for the wealthy, but eventually develop a large travel market suitable for the middle class with various price points and experience-related demands. Space explo-ration has developed from being only a government-funded activity to involving the military and commercial enterprises, establishing a need for international engagement in enhancing space sustainability. So far, the cooperation between governments and the private sector has tended to be economy based, leaving environmental issues somewhat voluntary (Duval & Hall, 2015; Webber, 2013).

In addition to their potentially lucrative economic potential, space-related activities have always played a major role politically, being espe-cially evident during the Cold War space race between the United States and the Soviet Union. The global space sector has already presented a good example for mutual international cooperation due to the high costs of space operations, but the increased role of the new space economy in current space activities has created an increased demand to avoid the former 'Wild West' attitude and ownership, for more detailed space legislation.

A strong human-oriented tradition is evident in the existing space law, and still a reluctance to see the space environment as worthy of protecting for its own sake, but as an economic resource for human utilisation (Viikari, 2007). In terms of sustainable operation, the dangers of space activities may not seem as imminent as those associated with similar problems on Earth, especially in the current megatrend of environmental protection. The increased commercial utilisation of outer space may have negative effects on the environment if mutual sustainably oriented global legislation is not soon thoroughly planned and activated to bind all the new space economy operators. The attempts to globally regulate the various environmental risks of space activities, such as the impact of space tourism, however, continue to appear relatively modest.

7 Visions of Sustainable Space Tourism

> They say any landing you can walk away from is a good one.
> Alan Shepard, Apollo 14 astronaut, 2020

Introduction

This last chapter looks at empirical findings from several interviews and an online public survey conducted between 2018 and 2019 to gather both the professional point of view as well as the opinions of the general public towards the new era of space tourism. In particular, it concentrates on the issue of sustainability and what the future may bring. All the interviews and the survey took place in Finland, as it is the native country of the author and also because it is a good representation of an industrialised Western nation, with a high standard of education and awareness of current global issues. Finland also became a new space nation in 2017 after Aalto University's Aalto 1 satellite was launched to measure electrons hitting the upper atmosphere (TEM, 2020).

The qualitative data was attained via three different sources: (1) the author's in-depth interviews with five professionals: a futurist, a politician, a space entrepreneur, a space scientist and a professor of space technology; (2) a professional Delphi panel (10 participants) run by the author, with participants from different professions related to space tourism: space engineers, tourism specialists, futurists, professors in space exploration and space tourism industry representatives; (3) a Webpropol online public survey created by the author (132 participants) targeting a university-educated adult population, on the open question of 'What kind of future space tourism would be sustainable?'. All participants' visions of the future of space tourism were treated anonymously. In the text that follows, individual quotes are attributed only to the profession, age or panel participation of the individual.

This chapter adds to the earlier discussion of sustainable space tourism themes: the human body; virtual reality (VR); space planning and legislation; technological innovation; and space colonies. The research participants' visions for the future enrich the context and the latest space tourism updates, especially from the Finnish media, show current ways of thinking in the country. It also offers insight from international sources, and shares the public's image of what sustainable space tourism looks like.

The Human Body

Previously, certain predictions have been made of crewed space missions operated by government organisations with the aim of developing the necessary technologies to carry humans safely to space and back. There are, however, two further motivations: one to learn about the space environments and their effects on the human body, and the other the exploration of space (Chhanivara, 2019).

Space is naturally a physically challenging environment for humans and a great deal of new technologically advanced innovation will be required before it can be inhabited. Some examples of how hard it is to colonise areas with harsh natural conditions can already be witnessed on remote corners of the Earth. The South and North Poles were conquered just over a century ago yet remain largely uninhabited – even though they have the same gravity as the rest of the Earth and greater potential than space for the transportation of goods such as food.

> Space is an unpleasant environment for the human body, and so pioneer space tourists will not travel to relax but to perform. Astronauts train for many years, and the handful of wealthy tourists who have already visited space spend numerous months in physical training before travelling. Mass space tourists will not have such time. The physical consequences for those without previous physical training remain unknown. (Space Technology Professor, 2018)

Even though there is some knowledge of how the human body performs in space conditions, the impact and long-term consequences of, for example, tourists travelling to Mars, remain unknown. There could even be positive impacts not yet discovered and the space environment could become a new treatment and health tourism option, especially once transportation becomes better suited to people with health issues.

> Some bed-rest studies simulating blood flow patterns in zero gravity with regard to cardiovascular diseases have demonstrated that it might be beneficial to be in zero gravity for an extended period of time. Blood pressure drops to zero and the heart can pump more easily – of benefit for people with heart diseases. The almost complete absence of gravity pushing on the intervertebral discs in the spine would also be beneficial for people with back problems, as these discs would be allowed to regenerate. (Space Scientist, 2018)

Eventually, if humans do start to live in space, evolution may begin to change the human physique to become more suitable for space habitation. The future human body may evolve through gene enhancement or even technical implants created through robotics and genetic engineering to give the body a transhuman upgrade. It is possible that genetically enhanced humans, such as bionic people, could shape the line of human development to adapt to space conditions. Future generations of humans may even be able to take on a variety of shapes and sizes dependent on their host planet and the conditions that best support the survival of *Homo sapiens* (Gallego, 2018).

> The health risks associated with space travel are quite high, so it is likely that a potential space traveller will encounter health issues when planning a flight. This is where potential tourists probably fall into two categories, those who do not want to go ahead with it, and those who nevertheless want to go into space. For the former, virtual experience is a good option, and also for those who can't afford the ticket, so this category may be higher. (Panellist, 2019)

Despite the potential health risks, there are already companies with almost utopian-sounding missions with no guarantee of actual success, such as SpaceBorn United, with a vision to organise and design missions where pregnant women could give birth in orbit. Presented at the Space and Science Congress in Germany in 2019, the company's visions include having the first baby born in space as soon as 2031, depending on funding and developments in the space tourism sector. The aim would not be to have the whole pregnancy take place in space, but instead a 24–36 hour mission for the labour. SpaceBorn United also focuses on embryo development and conception in space: 'If that sector is going to accelerate in the way it is doing right now, there will be markets for very wealthy people, who are not prepared to do three months military training. There will be spacecraft that are very comfortable for those people, it depends on the risk you are willing to take' (Dr Edelbroek, CEO, SpaceBorn United in Space and Science Congress, Germany, 2019).

Following an initial period of excitement, space tourism is likely to experience a sharp decline owing to its unpleasant conditions for humans. To allow the development of the space industry to advance, space hotels or stations providing gravity must be guaranteed, as well as 'pit stop' support for regular colonial travel on the surface of the Moon or Mars. It would also be advisable to look into the asteroid zone instead of Mars, which as a planet is very harsh on the human body. Close to zero gravity will make human bones brittle and cause muscle to disappear. The path of asteroids should somehow be controlled also, as they present a real danger given that an asteroid can hit the Earth. There are mineral reserves in the asteroid zone and mining these, using artificial inhabitants such as robots, could help in the construction of large circulating capsules for future human habitation (Peltoniemi, 2018).

Providing culinary experiences and other cultural activities, such as music concerts, in the same format as on Earth will be difficult in space. For example, if Mars were to become inhabited, new solutions for producing tasty food, or perhaps even new culinary discoveries that only the space environment could authentically offer will be required.

> The space environment raises questions about how to eat properly and enjoy music and other sensory cultural activities, whilst remaining in a non-gravity-based environment. (Space Entrepreneur, 2019)

Some research has been carried out on how primitive cyanobacteria interact with Mars at low atmospheric pressure. Cyanobacteria are sustainable single-cell organisms that could be used in food production and also produce oxygen, bind carbon dioxide and produce biomass. Some species can be used directly as human food, making cyanobacteria one of the most likely basic elements of space food, as they can be produced under very modest conditions (Peltoniemi, 2018).

China's Chang'e-4 creditor landed on the back of the Moon in early 2019 with the objective of exploring the distal side of the geology of the Moon. China managed to get seeds to germinate on the Moon's surface and the Chinese Space Administration later announced that the cotton seeds had sprouted, making this the first time that humankind has managed to grow biological material on the Moon. Before this latest advancement, the International Space Station had also successfully cultivated plants in a weightless state. The successful cultivation of food will make it easier to prepare for future long-distance space flights, such as those to Mars with a travel time of about two years. If astronauts are able to grow their own food, there is no need to take a full supply of food on board, and the Moon could be used as a stop-off point on the way to Mars (Nurminen, 2019).

Studies conducted by ESA (2019b) have looked at the impact of human hibernation on space missions. Based on sending six humans on a five-year mission to Mars, the study suggests that using hibernation would allow the mass of the spacecraft to be reduced by a third and the number of consumables cut by a similar amount. The process of hibernation would involve the astronauts or space tourists acting similarly to bears, being placed into individual sleeping pods on the craft that would double as living quarters after awakening. The soft-shell pods would maintain a cool, dark, environment to keep the crew in hibernation for the 180-day passage to and from Mars, followed by a 21-day recuperation phase. Passengers should not experience any bone or muscle loss even though they could be subject to cosmic radiation if the quarters were not protected by water tanks or other materials. However, hibernation might provide some enhanced radiation protection in itself (ESA, 2019b).

Virtual Reality

While the roots of VR can be traced back to the 1860s, when 360-degree art through panoramic murals began to appear, the technology is still in its early stages and existing VR gadgets are not yet able to completely replace the real-life experience. However, even though competition in the space tourism industry and rapid technological advancement are forecast to reduce the cost of space travel flights, it may easily take another 20 years, and in the meantime VR technology will bring a realistic space tourism experience at an affordable price.

> Virtual travel will represent the most likely means of 'visiting' space in the future, given that this environment is exceptionally hostile to humans. Although space tourism is physical, it can also be facilitated by nanorobots or avatars. Indeed, nanorobots may advance sci-fi storylines before humans are physically capable of such travel. (Politician, 2018)

A number of major consumer electronic companies on the market are offering VR headsets, advance image processing and sensing and robotic technologies to consumers (Chhanivara, 2019). However, in the space tourism VR experience, major obstacles remain concerning the transmission of content.

> I believe that VR technology can provide a taste for space travel and thus serve as a marketing tool. Of course, first of all, with high airfares, the general public will have to settle for the virtual equivalent of space travel. However, when ticket prices drop and the right space travel opportunity opens up to the general public, people will choose it. (Panellist, 2019)

Already there are VR innovations to help humans get closer to space. One example is the University of Surrey (UK) using VR to launch their very own trip into the stratosphere, with commercially available and affordable VR technology a tool to open up opportunities to the public to experience space travel at a competitive price. The idea was to create a virtual space ride so that the virtual traveller could sit in an armchair hooked up with a headset while the operating robot is on Mars. The video footage would be collected by cameras carried to the edge of space (20 km) by a high altitude balloon to create a virtual immersive journey (Chhanivara, 2019).

> Virtual tourism is growing rapidly as technologies evolve and become cheaper. The development will be fast to a holistic and multi-sensory experience. Due to the environmental impacts of space travel, its costs and risks, I see this option as highly likely and desirable. There was

already a time when the Internet was believed to kill physical tourism, but all scientists agree that the Internet has actually played a huge role in the growth and accessibility of tourism. Marketing and sales in particular have been able to take advantage of these opportunities and seeing places on screen has not replaced the physical experience, rather increased the need for it. However, studies also highlight the possibility that sufficiently realistic and interactive virtual experiences could already replace tourism. Of course, it may happen that virtual experience acts as a great marketing tool – and probably at least some of us want a real space trip. (Panellist, 2019)

The children of today have a realistic chance of becoming real space travellers at some point in their lives and are growing up around many space-related products. NASA even rhetorically promotes its space tourism posters as a part of virtual solar system exploration: 'Take a virtual trip to alien worlds, and maybe even plaster your living room with a new, futuristic space tourism poster created by a team of visual strategists' (NASA, 2019e).

Virtual reality has its benefits, and will surely be a way of experiencing, learning more about, and advertising space travel. Though, I doubt that anybody set on doing an actual space journey (and having the means to do it) would be satisfied with the virtual experience alone. People are unfortunately way too egoistic to just keep that experience as their single option – it is relatively well known how polluting current day air travel is, and still journeys are sold, and people fly for leisure purposes. (Panellist, 2019)

With the assistance of an advanced VR set, children could experience what it feels like to see the Earth from above and fly into space, and grow up in an era when space tourism is a reality: 'SpaceVR's innovative and immersive technology sails you into seas of stars as you float weightlessly experiencing the awe of Earth. Discover the planetary perspective that only a few astronauts have with the highest resolution VR camera ever launched in the space of the history of humanity' (SpaceVR, 2019).

Already there are so-called virtual trips to Mars. Nowadays, a school bus is transformed into a MARS bus: as it drives through New York's streets, Mars scenery is visible. As such, I do not see virtual tourism in space as a form of tourism, but instead, for example, as a form of teaching or as a new kind of 'non-fiction book'. (Panellist, 2019)

Combining the astronomy experience with travel, even the virtual kind, could foster an understanding of how the world is connected to the universe and, in terms of cultural sustainability, create a sense of protectiveness towards the Earth (Wilson, 2018). Experiencing moments of awe as they learn about our world could be life-changing for children and even inspire them to help our world in their own, unique way. There has been research concluding that feelings of awe produce a greater sense of well-being and a tendency towards altruism, making people more willing to volunteer to help others as well as more patient and less materialistic, thereby supporting the cause of sustainability (Rudd *et al.*, 2012).

Space Planning and Legislation

A valid concern exists around the planning and policy implementation of future space tourism, especially with respect to the impact of future agreements developed by politicians and the private sector and the importance of mutual accords geared towards leaving some parts of space completely untouched by tourism (Duval & Hall, 2015).

> Sustainability is a big trend in the tourism industry, and many are already ashamed of their own flying. Space travel will be very 'embarrassing' in some countries or social circles because of its unsustainable nature. But there are always enough people in the world who are not interested in it and commercial players who want to make money from it. One way or the other, compared to the overall demand for tourism, space travel has been a micro-activity for decades. But how do you know – if prices go down enough. (Panellist, 2019)

The new era in space exploration has brought about a completely new commercial market sector. In the past, considerable expertise, scientific research and economic capital were required to become a space nation, without any guarantee of direct rewards.

> Today a complete rocket can be purchased privately for a cost of approximately $100 million from the previous national owner. Private launch providers selling space inside such rockets, and customers, including small satellite companies, can already be found. The cost of the space inside a rocket is shared; hence, even academic institutions with relatively small budgets have been able to test small measuring satellites in a real space environment. (Space Scientist, 2018)

> Numerous possibilities exist to use space economically. For example, it would be more efficient to run solar panels in space than on the ground because UV light energy is more abundant in space. Plans have already been devised to locate solar panel stations in orbit and beam down energy using microwaves. (Space Technology Professor, 2018)

However, unresolved issues with asteroids and other debris have hitherto prevented such activity from taking place. Aside from space tourists simply experiencing the space environment, other actors outside of the tourism industry may initiate mass space travel in the future. For instance, asteroids are full of precious metals such as diamonds and platinum, and thus space travel may become work related (Ross, 2002).

> The development of synergies or by-products could be facilitated, including adventure products and services that do not yet exist. Indeed, space tourism could pioneer the creation of such services, which may be provided by other industries in the future. (Space Entrepreneur, 2018)

> If the sharing economy becomes an enduring global megatrend, the private space tourism sector will benefit, if not in actual existence, then at least in shared physical hubs between private sector parties, the military or space nations. (Futurist, 2018)

At present, considerable foresight and capital are required to forecast the economic potential of space tourism. With the assistance of governmental planning, companies should mutually plan, invest and share operational facilities to reduce the economic costs involved. Eventually, the expense will help determine whether space tourism can become a mass product or whether it will be practised only by the elite.

> It might be possible for an international body to ban space travel altogether, but I don't think so. Instead, I would hope that more sustainable forms of space travel could be produced and tested, moving away from space travel as a hobby for a small elite. This could also provide climate compensation for space travel, which could try to repair an uncomfortable conscience. (Panellist, 2019)

> I would not deny rich people financing space travel development but would – instead – advise them to seek more urgent investment opportunities that focus on equal access to basic human needs of services and products. The success of space tourism would also give hope to those people that do not care about sustainability, but count on having money for themselves in order to escape possible inconveniences caused by climate change. (Panellist, 2019)

As space becomes commercialised, mutual global legislation will be required to supervise future business and trading. The Outer Space Treaty (1967) corresponds to maritime law and constitutes the only legislation in existence for common space responsibilities. Otherwise, space is equivalent to the former 'Wild West' with a 'first come first served' attitude and approach to rights.

> For example, a future hotel or mining colony located on the Moon will raise questions as to whether the Moon should be regarded as a separate

state as well as which country's (or countries') legislation would be legally binding there. If the development of a mutual global space law proves too challenging, space should be treated similarly to Antarctica, which is divided into sectors belonging to different nations. (Space Scientist, 2018)

By 2040, global sustainability issues will prevail. In this respect, it is quite certain that the space industry is also 'subordinated' to sustainability goals. (Panellist, 2019)

Space is no longer only entered by countries with space programmes, but also various private companies working as subcontractors or even commercial contractors such as satellite operators. Furthermore, the military (worldwide) has diverse interests in space and is constantly developing technologies from which corporate space tourism companies may benefit (Insinna, 2018). However, there is currently a risk that a 'no man's land' attitude will arise, with the potential to cause ownership wars and jeopardise such synergistic operations.

Sustainable development is important of course, but at least equally important is to make sure that conquering space does not lead to a war between humans and nations. (Panellist, 2019)

Once the private space tourism industry has developed infrastructure that can run independently, government sponsorship and funding is likely to cease. Without the support of a governmental legislative framework and preventative space law, private companies will face new problems related to owning and immaterial rights. (Futurist, 2018)

In addition to ownership, a further major legislative concern is around responsibility for space debris. Even a small particle of debris from a former satellite or similar may prevent a safe landing back to Earth. This type of safety threat could seriously harm the entire space tourism industry.

The use of space is bound to be regulated, otherwise we can no longer keep satellites in orbit. That would be a really big problem for future civilisation. Space technology is needed to solve almost all global problems. (Panellist, 2019).

Preventative space legislation should make such debris collectors compulsory in the future to ensure safe access to space and protection of the space environment in general.

I would say that space legislation is particularly required because of the satellite activity already in place and the future of material promotion. Satellites have produced vast amounts of space debris, which will make future space activities more difficult if something is not done. I believe that, rather than space tourism, commercial activity will focus on the promotion of asteroids, which in itself requires its own laws. (Panellist, 2019)

Technological Innovations

Space has become increasingly important economically and politically for many formerly non-active countries, providing more tools for scientific research and economic investment opportunities, with examples being independent navigation systems such as American GPS, European Galileo and Chinese BeiDou. Many small-sized countries are currently building their own space programmes. The less than a decade old start-up company, Planet, is now operating the world's largest satellite constellation, with over 200 satellites that can produce an image almost every three or four hours. The fact that a space-related start-up company, originally founded by three students, can reach such a high operational level in only a few years, receiving hundreds of millions of dollars in funding, has also been a completely new phenomenon in the space industry (Wallius, 2018).

As NASA stopped its Space Shuttle programme in the 2010s, cooperation with the private sector was initiated and private companies managed to develop rocket technology on a schedule that no one could have believed possible. One of the reasons for such speedy progress was that the technology advanced very rapidly and was able to lead to miniaturisation. Satellites are so small that a hundred can be packed in a rocket, resulting in a drop in price. There is power in this larger volume as the constellations of hundreds of satellites generate up-to-date information on congestion, harbours and, for example, the ice situation in northern waters. Artificial intelligence is also required for screening images, and the next area of advancement for satellites is in telecommunications and the Internet of Things (Praaks, 2018).

Currently, over 4500 satellites orbit the Earth, though only about 1000 of them are still working, leaving the rest to become space debris, endangering new satellites as well as impacting on safety factors for space tourism (Heikkilä, 2018). The European Space Agency has estimated that there are more than 29,000 debris particles over 10 centimetres in size and more than 166 million particles less than 10 centimetres. Since the space environment does not have friction, a particle may continue to be in the Earth's orbit over hundreds of years, moving at the same original speed, over seven kilometres per second, and even a debris particle weighing 1 gram can cause the same damage as that of a 300-gram particle hitting a car's windscreen on a motorway (Palmroth, 2018).

New technical solutions are therefore needed, such as space debris collectors or a technical tool attached to a satellite to bring it down to Earth safely, to ensure that the situation does not accelerate to become unsustainable, and also to guarantee the safety and functional ability of both future satellites as well as space tourism vehicles, especially those with large viewing windows (such as Blue Origin's). Future satellites should also be designed sustainably to have a longer life expectancy as well as an ability to be controlled at the end of their 'living path' to be escorted back to Earth.

> In space jumping tourism, twisted scrap cannot occur. In orbit tourism in principle it can, but it is easily avoided. However, there is an existing problem of orbital rotation caused by the satellite industry and, if not addressed in time, it could threaten the realisation of orbital tourism, in addition to threatening satellites in the 500–1,200 kilometre altitude range. If the orbital debris problem is not treated, rendering LEO orbit unusable, space tourism can go further, for example, to Lagrange points. (So much orbital waste is not going to be flown through, inactivity applies to longer stays at dangerous level). (Panellist, 2019)

According to Daniel Lockney, the chief of NASA's technology development programme, NASA's visions for the near future are first to return astronauts to the Moon by 2024 and then to continue to Mars or asteroids, with the Moon acting as a pit stop for the journey. NASA plans to develop and test reasonably priced and reusable technical solutions, such as dust prevention inside space vehicles, first on the surface of the Moon and later to be utilised on the Mars Mission (Kivimäki, 2019). By 2022, the United States plans to build a new space station called Gateway, which will be smaller than the International Space Station, located near the Moon, to act as a permanent base for further space exploration (Paukku, 2019b).

Currently, there is a boom in private businesses targeting space for opportunities that will have enormous future economic impacts. A private company, Moon Express, was granted permission from the federal authorities of the United States to use a private mining robot to extract the surface of the Moon, while it was contending in the Lunar XPRIZE competition, which finished without a result as the participants did not make it to the Moon by the given deadline. Other private companies, such as Planetary Resources and Deep Mine Industries, are also aiming to start pioneering mining for valuable minerals on the Moon as soon as the commercial space industry advances (Paukku, 2019a).

Currently, China is at the forefront of the exploration of the Moon as it recently landed there with a descent that could, in principle, even be possible for crewed flights. There are, however, questions concerning the impact of the magnetic field which should first be properly explored with robotic bullets and landers instead of humans (Wallius, 2018). Russia launched a human-sized humanoid robot, Fedor, into space in August 2019, to spend 10 days on the International Space Station to learn how to assist astronauts in technical support tasks such as using cables, mechanical tools and fire extinguishers. Since 2011, two other humanoid robots from the United States and Japan have been in space (Sulasma, 2019).

The electric solar wind sail, an idea first conceptualised by Pekka Janhunen of the Finnish Meteorological Institute, is a proposed form of spacecraft propulsion using the dynamic pressure of solar wind as a source of thrust. It creates a virtual sail by using small wires to form an electrical field that deflects solar wind protons and extracts their momentum. A full-size sail would have 50–100 straightened tethers with a length of about 20 kilometres each. The electric sail is also able to continue to accelerate at greater distances from the sun, still developing thrust as it cruises towards the outer planets. Inside the planetary magnetosphere, the electric wind sail can perform a brake function, allowing for deorbiting of the satellites, which means that interstellar ships approaching the sun may use a solar wind sail for braking (Janhunen, 2004).

> Some existing technologies may help remove debris from orbit without requiring a propellant. An electric sail (1–2 kilometres in length) can be attached to a small satellite and, once the decision is taken for it to leave orbit, the wire is unspooled from the satellite and the solar wind pushes it into outer space. (Space Scientist, 2018)

Another space sail research and engineering project is by Mark Zuckerberg and Yuri Milner, called Breakthrough Starshot, which aims to prove the concept with a fleet of light sail spacecraft capable of making the journey to the Alpha Centauri Star system over four light-years away. The Breakthrough Starshot operates by first raising a large sail to orbit the Earth, then releasing a thousand smaller sails about one metre in size, with small radios in the middle. These star sails are then accelerated to one-fifth the speed of light with the assistance of large lasers. The purpose of the project is to advance sustainable technology development, as a journey to Mars would only take a month using this sailing technique, and also to push space debris out of the orbit (Paukku, 2019c).

Recent technological innovations also include a steam balloon, initially filled with hot steam on the ground and then released at a high altitude where the air is thin, to facilitate satellite launches (FMI, 2019). Launching from a high altitude will reduce air drag, thereby improving efficiency, and after descending the emptied balloon may be collected for reuse. Janhunen (2019) envisions that it will also be feasible to utilise such balloon types within the space tourism industry. A tourism capsule can be raised to the stratosphere with the steam balloon and dropped so that the passengers experience brief weightlessness before a parachuted descent. The motivation for using steam is that it is cheaper and also more sustainable than hydrogen and helium, and it is not flammable.

Space Colonies

There are many arguments for and against space colonisation, from Stephen Hawking's view that space flight and the colonisation of space are necessary for the future of humanity as otherwise the human race will become extinct within the next thousand years, to the idea of biosphere survival in the event of natural or human-made planetary-scale disaster, to the availability of additional space resources to enable the expansion of human society. Some concerns have been raised about the enhanced interest and exploitation by organisations that are already powerful, such as military and major economic institutions, which could exacerbate pre-existing detrimental processes such as economic inequality, environmental degradation and even war.

> In order to ensure human survival, such as in the case of a global catastrophe like a comet impact, it would be sensible and advantageous to begin developing space colonies. This will require advanced technological innovations, although some already exist in some capacity. The first permanent settlements will most likely be located on the Moon or Mars. Instead of just admiring the curvature of Earth, future space tourists may also travel to visit friends and relatives living in such settlements. (Space Entrepreneur, 2019)

> I find space colonies to be neither probable, nor desirable – we have one Earth, with tremendous opportunities, but we are gradually eroding away this in our pursuit for the next big thing, the next growth, the next... something. The dream, or scenario, of creating colonies outside Earth 'in case of' is a lame excuse that is trying to create a utility value for a matter that hasn't got one. It is comparable with US gun laws that allow people to be armed with automatic assault rifles 'in case of being attacked', even though they are almost useless as defence weapons (unless attacked by a hoard of enemies). For the greater good these 'colonies' would anyhow mean very little. If Earth went through a global catastrophe then there is no place for 7.2 billion people – and no other environment replenishes itself like Earth. It is therefore a 'pie in the sky', literally and figuratively, to wish for space colonies – whilst simultaneously making the Earth that exists in front of our eyes less and less liveable. (Panellist, 2019)

For the first time in the history of space exploration, scientists have recently been able to measure the seasonal changes in the gases that fill the air directly above the surface of Gale Crater on Mars and, as a result, have noticed that the oxygen levels of the Martian atmosphere follow a seasonal pattern. This kind of environmental phenomenon has not been witnessed before and is under further investigation with the new sample analysis at Mars tool (SAM) measurements replacing the former NASA

Viking landers that arrived on the planet in 1976 (NASA, 2019m). These types of scientific discoveries provide essential new knowledge about future living conditions for humans settling on the Red Planet.

> As social human beings, there will soon be a push towards having companionship in space and being the first human to establish a colony will be very attractive to people who can afford it. Setting up a colony there would be a major scientific advancement, that's why it would be good. And yes, it is possible that after facing a tremendous catastrophe, it might be the only way to save humankind. (Panellist, 2019)

> 'Second Life' platform studies have already demonstrated that, even though people are afforded the freedom to use limitless imagination in virtual reality, they tend to carefully copy real-life structures. (Futurist, 2018)

China is taking steps towards further exploration as it plans to launch a Mars probe in 2020 that will orbit Mars and then land to explore the surface, with the mission to test technologies that China could use in future ventures such as delivering supplies (Yamada, 2016). NASA's Mars 2020 all-terrain vehicle will land on the Jezero Crater in 2021 to collect any traces of life such as fossils and will use astrobiology as a tool to understand why Mars has changed from a planet with running water and an atmosphere to a planet hostile to life (Keränen, 2019a). If some key scientific breakthroughs are made, the information will be very beneficial for the development of future space colonies. However, the first colony from Earth may already exist on the surface of Mars, in the form of bacterium, as in the early days of space exploration the probes were not properly disinfected.

> The earliest time span estimated for initial settlements on the Moon is 20 years. The development is likely to commence with the erection of greenhouses to ensure food production and to test survival-related technologies. Within the next 100 years, self-sufficient colonies will exist on the Moon. Whether colonies will become viable on Mars depends on factors that are currently unknown, such as the impacts of long-term exposure to cosmic radiation while travelling. (Space Scientist, 2019)

As already summarised in the late 1970s by Bluth (1979) and still appears completely valid today, the colonisation of space will require vast amounts of devoted human capital in the realms of physics, research, finance and technology. It is therefore necessary to have a clear end goal to not unnecessarily waste the limited natural resources of the Earth. As the public investments needed for such space programmes are in the end collected by taxation, it certainly raises issues around fairness, such as who has the right to benefit from the existence of a colony in the case of a sudden emergency, such as a comet impact on the Earth. There is a valid ethical concern that the already powerful may become even more so, leading to economic inequality, social division and environmental degradation. Additionally, issues surrounding human dignity, morality and cultural differences will become apparent in efforts to meet the social needs of people from, most likely, various international backgrounds living in isolated space colonies.

Sustainable Future Space Tourism?

The aim of the open-ended question of an online survey of the public was to find out the respondents' opinions and visions when combining the familiar existing megatrend of sustainability with the more distant idea of future space tourism. Selective samples of the given answers are placed under four different age categories: 'under 30 years old', '30–40 years old', '41–50 years old' and 'over 50 years old' to show the viewpoints of different generations. The quotes include the gender as well as the current occupational status of the respondent to frame the context.

Under 30 years old

I imagine space tourism in which emissions are not too high from a potential launch. Although I do not know how much rocket launches will emit, I do not think emissions will decrease if space travel becomes a new way to travel. (Male, bachelor's degree, professional)

I don't see any clever reason why humans should also enter space. It's time to see how the former islands of paradise are ruined today and I don't want the same thing to happen to space before humans can even be proven to take care of the Earth. So no space travel. (Female, bachelor's degree, employee)

Minimising space debris is paramount. The durability and reuse of used equipment must also be a priority. Significant resources must be invested in the selection and development of potential propulsion. (Male, bachelor's degree, employee)

Environmentally friendly – no significant pollution. Such that most people could afford, with no division and possible sense of discrimination/inequality in humanity. (Male, bachelor's degree, student)

The most ecological and inexpensive implementation possible, possibly in a way that supports the well-being of the environment that people could use as an option instead of higher emission vehicles. (Female, bachelor's degree, student)

Ecological issues, such as concern over increased pollution and the creation of debris, appeared to be paramount within this age group. These respondents also questioned why people should enter space in the first place, as the most likely result will be the destruction of its environment. The answers given reflected the younger generation's anxiety about climate change and worries for the future of the Earth.

30–40 years old

> In my opinion, as long as mankind is mentally so low, we have no reason to go into space. This planet must first be balanced, and humanity must become much wiser before space travel makes any sense. (Male, master's degree, unemployed)

> I guess it's supposed to be there for quite some time, so it kind of balances the load? If a missile with a huge amount of fuel is shipped from here, they will be shipped less frequently, not every day, like the planes that fly to the Canary Islands. Let's stay in Jupiter longer than a long weekend. (Male, master's degree, employee)

> The best option for the environment is without space travel. Taking the existing space shuttle from Virgin Galactic to space requires a lot of fuel to burn, which is not good for the environment. Just wondering how many commercial flights can cover just one journey of a few people to space? In my opinion, just a few rich people should not be able to contaminate this planet as much as hundreds of normal people traveling by aircraft. Space traveling is not moral nor ecological. (Male, bachelor's degree, professional)

Concern over ethical and equality issues was dominant within this age group. It was suggested that to justify space travel, it should be last longer than just space jumps, to justify the emissions created.

41–50 years old

> Space travel always requires a great deal of energy and emissions, and it is not theoretically possible to carry it out ecologically. Fuel can of course be produced organically, but its use for space travel is always unecological (fuel should be used for other purposes if you want to be ecological). (Male, master's degree, senior professional)

> Science, promoting and/or educating a much larger number than the number of actual travellers about the universe (mass media, social media influencers, writers, institutional spokesmen etc.). (Female, master's degree, employee)

> Virtual? A virtual reality-driven, thought-provoking, travel experience? It's as if you feel like you're in space, even when you're on Earth. Developing such 'tourist experiences' through technology could be a good thing. As such, it is good for man to have a perspective on his own planet

and his existence from space. I remember hearing that the first pictures of the little blue Earth sparked an environmental movement. But space travel should not be a playground for the wealthy who want to experience 'an experience for those who have experienced it all'. Mass tourism, too, sounds bad, at least until you can travel to space under conditions of sustainable development. (Female, master's degree, senior professional)

No fossil fuels, virtual travelling from the couch at home. (Male, master's degree, professional)

Space travel sounds like a very distant alternative in my own life. The best would be unpolluted, unproductive of travel by-products. In general, any unjustified 'traveling' is useless, so the same applies here. (Female, master's degree, manager)

Concerns about emissions and the impact on the environment were highlighted within this age group. However, some solutions were suggested to justify space travel, such as spreading cultural awareness to larger audiences as well as experiencing a space journey virtually from the comfort of one's own home.

Over 50 years old

One that provides as much information, understanding and experience as possible about the operation of our own spacecraft in relation to phenomena elsewhere. For example, information on the historical environmental development of our two neighbouring planets gives us the best insight into how the Earth works and how we should proceed. The Moon colony would provide clear guidelines on how to maintain life in a closed system. And the Earth is a closed system except for the energy provided by the sun. And like Musk says, long-haul flights would be the best for the atmosphere through space. (Male, master's degree, pensioner)

Organised by international communities following the mutually agreed legislative framework and regulations. (Male, master's degree, manager)

The key issues within this age group concentrated on the necessity of forming valid global regulations as well as, similarly to the age group of 41–50 years old, an emphasis on educating people to understand planet Earth's fragility within the universe.

Conclusion

This chapter has provided a more thorough investigation of the five thematic areas of the human body; VR; space planning and legislation; technological innovations; and space colonies by utilising some of the findings of empirical in-depth interviews, a professional Delphi panel and a public survey conducted by the author during 2018–2019 for her PhD research process on sustainable space tourism.

Firstly, the human body section linked to the survival element of sustainable development of *Homo sapiens* and pointed out that the space environment is very challenging for and hostile to the human body, which will obviously impact human exploration of space. Due to advanced technology and even transhuman upgrades, the future human body may be better facilitated to meet the harsh space conditions and to create a human habitat in space. Secondly, the VR component related to the cultural element of sustainable development by noting that virtual travel will most likely become a popular future option to 'visit' space, offering a realistic, environmentally friendly and affordable experience accessible to most people, and creating a tool for education by generating a sense of protectiveness towards the Earth in the younger generation.

Next, the section on space planning and legislation linked to the economic level of sustainable development by examining the concern over practically non-existent global regulations on space legislation and planning policies, especially as the new space economy has brought in new operators, such as space tourism enterprises, with a 'Wild West' attitude. The topic of technological innovations linked to the operational level of sustainable development, as many smaller nations and start-up companies have been able to build their own space programmes due to the miniaturisation of space technology. This has provided more tools for scientific understanding and discoveries to enhance sustainability, such as the electric wind sail, which could clear space debris and use solar power as a source of energy for travel.

The fifth element, space colonies, related to the resource level of sustainable development, pointing out that colonising the Moon and Mars would be a major scientific advancement as it ultimately could save *Homo sapiens* from extinction as well as transform some of the industrial activities from Earth to other planets of the universe – to give our planet a new chance of environmental recovery.

Lastly, a common attitude garnered from public opinion about space tourism was that as long as there are increased emissions caused by space tourism launches, the most sustainable option for experiencing space tourism would be through VR experiences and other terrestrial activities. However, if the space tourism industry justifies its existence by providing long-lasting positive impacts for society, such as enhancing sustainable technological innovations that would also be suitable for the Earth, collectively joining together in debris collection in space with other new space operators and enriching the knowledge of our place in the universe especially in younger generations, then this new adventure tourism industry may start promoting itself as a sustainably-oriented one.

Concluding Words

We are currently living in an era of a new space race, during which the traditional aviation industry's plans for higher flying altitudes to shorten flight times between long-distance destinations and the actual space environment have lately become dominated by the operations of the new space economy. There is increased cooperation between governments and the private space industry sector, meaning that the success of civil aviation is a good economic model to follow for space industry development, instead of the previous government monopoly space agencies. Space tourism is a logical development of the ever-increasing distances that tourists travel and the exploration of the space environment by robotic or crewed missions is a natural extension of humankind's desire to explore our own planet (Williamson, 2003: 47). Space tourism could become the first large-scale commercial user of hydrogen, influencing its future use in aeroplanes and other ground transportation systems. Breaking the familiarity barrier may even engender a new golden age of rapid environmental knowledge and protection. The sustainability requirements in space are demanding and must be solved here on Earth before being practised in a space environment. The focus should be on sustainable science, achieved through understanding the psychology of new types of tourism behaviour as well as global environmental agreements.

The first chapter of this book introduced the historical milestones of space tourism, defined different types of activities belonging to this new tourism sector and explained the current status and elements of the space tourism industry. Space tourism will soon turn from science fiction into reality. The original excitement created by Hollywood productions was followed by decades of passivity, until the 2000s when a number of technological innovations enabled the space tourism industry to make rapid steps towards the first commercial space flights. The actively growing space tourism industry has three major operators with private and influential funding: Virgin Galactic, SpaceX and Blue Origin, competing for the status of pioneers of human space flights, and all originally benefiting from governmental initiatives such as NASA's Space Launch and the XPRIZE competition to prepare the technology for a reusable launch vehicle. As of early 2020, no suborbital space tourism has yet occurred, but in December 2018, Sir Richard Branson (founder of Virgin Galactic) was able to announce, 'Today, for the first time in history, a crewed spaceship built to carry private passengers reached space'.

Currently, space tourism options with the lowest amounts of travel, training and mission time are 'zero gravity flights', in which tourists experience the feeling of weightlessness, and 'edge of space' flights, in which the curvature of the Earth can be witnessed. Upcoming, but expensive, suborbital flights will require a proper space suit for viewing the Earth and future translunar cruises are envisaged to transport passengers to the Moon or space hotels nearby in lunar orbit. The pioneering space tourist must undergo various health checks to satisfy safety authorities and insurance companies. However, increasing the range of virtual reality options and other terrestrial forms of space tourism, such as watching the Northern Lights, will go some way to equalise this future adventure tourism sector, making an almost authentic experience available to a wider population.

The second chapter investigated the discourse of sustainability, focusing especially on the growing concern of climate change and the factors surrounding it. In the past decades, the issue of sustainable development and concerns about the impact of population growth and energy supplies were commonly recognised by individuals, organisations and governments, but the IPCC (2018) report still managed to shock the world with its pessimistic forecast of the Earth's future. In the worst-case scenario of the Earth becoming uninhabitable in the future, similar dynamics to those identified in relation to space flights may lead affluence to play a significant role in dictating who is empowered to escape the Earth and attain the benefits of residing in outer space (Spector & Higham, 2019a).

Due to the current global concerns around emission levels and climate change, there is obvious apprehension about environmental issues arising at the start of the commercial space tourism industry. Space tourism will involve a variety of risks that can be termed environmental and detrimental effects may impact on the Earth, atmosphere, space environment and human activities in these environments. Space vehicles will travel through the atmosphere, where discharge from rocket motors deteriorates the ozone layer, diminishing Earth's protection against ultraviolet radiation, and contributing to acid rain (Viikari, 2007). The particles of black soot emitted by spacecraft can impact the surface temperature of the Earth as the launch of a spacecraft may leave black carbon particles in the stratosphere for 10 years (Bradbury, 2010), and just 1000 space tourism-related launches could affect a global temperature increase of up to 1°C (Ross et al., 2010).

Peeters (2018) provided a critique that 1000 launches per year is a tiny number, placing a great responsibility on those few people benefiting from the launches, and Spector et al. (2017) suggest that greater attention to space tourism and commercial space flight is required in order to develop a coherent, long-term conceptualisation of the implications of modern mobility on sustainability.

The Arctic, with its remote, harsh living environment, could be used as an experimental example to mimic the impact of the infrastructural maintenance needed for securing the health and safety of space tourists in the space environment. It should be noted that weak signals indicate that space exploration is already showing signs of early colonisation through the formation of space military and such actions are resulting in more environmental debris in space. The desire of the world's powerful countries and companies to exploit the natural resources creates questions not only around the environmental impact on the untouched space environment, but also around creating global equality for those not yet able to access space.

The third chapter defined some of the elements of future forecasting and introduced new future forecasting models to be used as tools for forming future scenarios, such as space colonies. The creation of commercial space tourism started in the wave of technological hype after the millennium. Paradoxically, during its developmental phase the consumer trends in travel simultaneously shifted to more sustainable practices. In order to survive, the space tourism industry needs to respond to this megatrend as well as follow increasingly stringent international climate change agreements. Appropriate regulations to encourage businesses to use a combination of different advanced technologies, so that they can avoid a destructive environmental impact, are required.

With the assistance of the Futures Map (Kuusi *et al.*, 2015) and the Sustainable Future Planning Framework (Toivonen, 2017), a long-term conceptualisation of sustainable implications for space tourism may be created for future planning processes. There is a strong emphasis on all the actors involved in the commercial space tourism industry taking responsibility for their role in creating sustainable tourism and, for example, to join universally recognised compensation schemes from the beginning, easing the 'environmental conscience' of both the space tourism operator and the environmentally aware customer.

Similarly to the current forms of extreme adventure travel that can be costly to operate, space tourism will first be a travel activity or even a hobby for the rich elite only – the mass will, however, eventually follow. New phenomena often start with ideas that are almost styled on science fiction, followed by awareness among the elite. Popular awareness, government awareness and the enactment of new policies follow thereafter (Markley, 2011). Humanity is technologically close to the point where colonies could be built on the surface of the Moon and Mars by utilising materials from these locations. Various perspectives that support future colonisation include the survival of *Homo sapiens* if Earth becomes uninhabitable, and the motive of replacing some Earth-bound resources with those obtained in space. The pioneering space colonist could initially build and maintain solar panels that would be used to provide power on Earth, which is more efficient than collecting the sun's energy on the Earth's surface after it has passed through the atmosphere (Johnston, 2017). The weak signals suggest that travelling to near space is likely to become a trend in the future, especially if space tourism becomes cost-effective for the masses and conscientiously adheres to environmental and safety requirements. Currently, the space tourism industry appears to be aiming to reach a similar state to that of early aviation; however, customers may eventually simply purchase a ticket at a launching site and shoot up to space to visit friends living in a Moon settlement.

The fourth chapter explored the elements in tourism planning and analysed how space tourism and sustainable development planning link together. Existing models for sustainable development and planning in tourism such as Plog's (1974) and Butler's (1980) life cycle destination models can measure the number of tourists against the evolution of the visitor types or against the rejuvenation or decline of the destination.

Emerging space tourism has several links to sustainable development, but technological solutions could also be considered a waste of resources, and this separation and antagonism between space tourism and sustainable development need to be overcome (Fawkes, 2007). The links between the two can be found at five different levels: operations, cultural, economic, resources and human survival. However, there is also critical opposition against this kind of classification; Peeters (2018) claims that there are no arguments to believe that space tourism could be a part of sustainable development on Earth and this is because, by posing outer space migration as a serious option, it may distract policymakers from taking the necessary mitigation measures.

Actions to support sustainability, similar to the environmental actions currently practised within the tourism industry, could quite effortlessly be included in many stages of the space tourism industry's infrastructure and marketing, from the design of materials to the manufacture of tour-ists' spacesuits, to global participation in carbon compensation schemes. The risks caused by creating a 'non-sustainable' image of the future space tourism industry must be mitigated from the start by rigorously introduc-ing sustainability into different operations and systems and clarifying the case that space will also be a vital part of sustainability.

Increased educational awareness and feelings of awe towards Earth, such as those already described by astronauts, may bring many positive consequences and create economic prosperity from money donated to different environmental protection schemes on Earth, and also bring space tourism-related employment opportunities to local communities. Satellites that assist daily activities on Earth, enhanced by the technologi-cal progress in reusable innovations in the private space tourism industry, can also speed up the response to natural disasters with accurate data supporting rescue operations, and monitor the impact of climate change more accurately. For the most optimistic or imaginative, space tourism could become a distant form of 'nature tourism', as seeing the Earth from afar might enhance people's desire to protect the only planet that can offer suitable living conditions for humans. It may even end up with a place on the list of sustainable tourism options, like ecotourism, volunteer tourism and agritourism (Peeters, 2018).

The shortage of the Earth's resources has been used to justify many natural resource wars between different nations and forecasts of population growth may enhance the fear even more. In the most optimistic of visions, utilising natural resources from space could be the new alternative to ensure more equal standards of living for everyone on Earth. It may even secure world peace, and the development of space tourism with its forecasted reduction in price could ensure continuity of such progress. In the future, outer space will play a central role in determining the long-term sustainability and even the survival of *Homo sapiens*. Considering the ethics of space tourism requires us to reflect upon the role of mobility in facilitating our presence beyond the biosphere (Spector *et al.*, 2017).

The fifth chapter examined space tourism's logical appearance in the human era of the Anthropocene as a result of behavioural changes in society, and explored some ethical considerations towards space commercialisation. The Anthropocene has altered the world beyond anything in its previous geological timescale. Recently, technological innovations, such as the creation of virtual reality, artificial intelligence and robotisation, have shifted the focus away from human actions, to non-human ways of operating, creating the beginning of the Post-Anthropocene. The first stage of space exploration has been accomplished with the technical support of robots and satellites, which also support the rise in human individualism and adventure seeking on Earth.

Space tourism may be considered the ultimate luxury tourism experience; however, at the same time, the concept of luxury has moved from the physical to the more emotive appeal, and space tourism is also a metaphor for understanding that change (Yeoman, 2008). Due to the current environmentally cautious megatrend, there is a new need for a worldwide restructure, to change the emphasis from the increased hording of material goods, easily leading to global ethical apathy especially among the rich of the world, towards more environmentally protective and further-reaching solutions. Examples include giving more emphasis to education to teach children about ethical aspects, as the next generation will most likely be the first to travel and work off the planet Earth, and ensuring that private corporations desiring to conduct business in space first have a 'space experience' to fully understand the functionality of their new business environment.

It is also important to consider social sustainability, relating to tourism degrowth and unequal development, in the creation of a new form of tourism. As previous tourism studies indicate, despite decades of concentrated effort to achieve sustainable development, social inequality is still present. There needs to be more critical consideration of the different aspects of global social fairness in relation to the space environment and opportunities for the residents of Earth to avoid ending up with an uncontrolled development parallel where the powerful dictate the rest of the world, in both economic and social elite dynamics of concentrated power and action.

The continued expansion and development of space also impacts on local communities, especially near new spaceports, as the case study of the United Kingdom indicated. The impacts may be both positive and negative, bringing more employment and economic opportunities, but also increasing many forms of environmental pollution. A sound governmental legislative framework, such as the UK's new Space Industry Act, will be a future necessity to ensure that there is a sustainable balance in the space tourism industry between the aspirations of private companies and the wishes of the local community in order to minimise some of the negative impacts already experienced elsewhere.

The sixth chapter explored the elements of the new space economy, particularly in terms of private space sector developments in reusability, and analysed the future economic forecast, including the cost structure and market demand for potential future space tourists. Space tourism is forecast to become an economically lucrative business, a $3 billion market by 2030, and superfast travel using outer space, another spin-off enhanced by space tourism, could be a $20 billion market and compete with long-distance airline flights (UBS, 2019). In 2017, $4 billion of private money was invested in space-related projects, almost entirely in the United States, despite space tourism not yet even operating.

It can be concluded that there was an unnecessary decades-long delay in developing public space travel, despite the basic technology having been present since the 1970s. This contributed to the delay in the creation of new technologically led industries, thereby constraining economic growth and putting a strain on Earth's own limited natural resources, causing more environmental damage for the entire planet. In 1998, NASA's feasibility study highlighted some positive outcomes of potential future space tourism, acknowledging its revolutionary political, social and economic potential.

For a long time, one of the key challenges in the development of space tourism was finding economic funding. However, after the millennium, Silicon Valley-based private sector billionaires such as Elon Musk (SpaceX) and Jeff Bezos (Blue Origin), and Richard Branson (Virgin Galactic) in the UK, established a new technologically driven private space race by introducing their own private space tourism companies that target the public. This immediately raised criticism and questions about the ethical synthesis of influential private sector commerce and publicly funded infrastructure, as it brought concerns regarding policy and the concentration of power (Tett, 2018). A similar concern also relates to the backgrounds of the pioneering space tourists; the handful of tourists paying over $20 million for their trip to the International Space Station came from already privileged segments of society and the soon to emerge low orbital 'space jumps' are also only affordable for the wealthy elite: will the understanding of an authentic space environment be a new separating power of knowledge, deepening the gap between the rich and the rest even more?

There has been some, although still limited, global space tourism market research highlighting that the public is interested in the idea of experiencing space tourism, and concluding that the commercialisation of space in terms of tourism could have many synergies, creating positive effects for the business, as without space tourism the rapid development of reusable and reliable low-cost launch vehicles could not be guaranteed and the goal of affordable access to space could not be achieved.

The importance of the lack of coherent space legislation in space tourism and the need for new global regulations for the space environment were also addressed alongside legally binding ethical and safety considerations. Traditionally, the rationale for space flight has followed five major themes for justifying the large-scale operational agenda: scientific discovery and understanding; national security and military applications; economic competitiveness and commercial applications; human destiny/survival; and national prestige/geopolitics (Launius, 2019: 27). Spector and Higham (2019: 6) warn that the present rationale for private industry space involves some risks: 'the current human pathway beyond Earth's biosphere is being driven by the interests of "space billionaires" who enjoy international legal regimes and natural regulatory systems that are highly conductive to private sector space developments and under such circumstances the democratisation of space travel and any benefits arising therefrom appears unlikely'.

Sustainable tourism innovation, including the adaptation and mitigation of climate and other negative environmental change, needs to be better understood. Governance, institutions, cultures and resources all serve to influence the actions of actors in space tourism companies (Gössling et al., 2009). In the long term and at global scales the space tourism industry needs to alter its structure in response to legislative and commercial pressure. It is important to acknowledge the significance of natural politics and decision-making to evaluate their environmental, social and cultural impacts and the responsibility of policymakers is paramount because high costs tend to create a barrier to sustainable actions. The legislation and management plans for space tourism are just forming, hence providing a perfect opportunity to design the rules and practices to meet sustainable criteria.

The increased utilisation of outer space means that there are increasing environmental threats, such as increased emissions on Earth and more space debris in outer space, and it has become obvious that the effective management of environmental problems related to space activities has not been assisted by specific sustainability-oriented international space law (Viikari, 2007). Currently, space is subject to a 'first come first served Wild West' attitude as the only existing Outer Space Treaty dates back to the 1960s and is not up to date with the situation created by advanced technological solutions.

There is a strong need for new global regulations concentrating especially on the new space economy activities, such as space tourism, and defining the ethical boundaries within which each sector is bound to operate. For example, in addition to limiting and regulating human activity in outer space, parts of space could be fully preserved from human intervention. Plans by the United States to reform their oversight of space development and regulations for commercial space flight launch and re-entry operations are currently being developed, possibly simply motivated by the fact that the country may soon be eclipsed by China in space exploration (Tett, 2018). However, under the climate change debate there should be a strong mutual willingness among the nations of the world to aim for a new effective space law to ensure sustainable continuity for future generations both on Earth and in space.

The last chapter explored some of the empirical findings, collected by the author with the assistance of in-depth interviews, a professional Delphi panel and a public survey on sustainable space tourism. The key themes, enriched with the findings related to the human body, virtual reality, space planning and legislation, technological innovations and space colonies, and some public opinions were shared to describe what sustainable space tourism means to them.

Despite the space environment being extremely hostile to the human body, the chance to experience the Earth from beyond, where the world is shown as one united place, may be a sufficient motive for people to decide to expose their bodies to space. As the technology advances, new space-related company announcements are likely to increase, some targeting the improvement of human bodily functions in space, while others concentrate on the production of space nutrition for longer space stays. With time, this may shift the focus of discourse about the human body in space away from the elements of danger and more towards the idea of leisure. Virtual reality technologies can change the way we interact in the future and provide a practical solution for many without the necessary funds or the necessary physical or age standards to have an authentic experience. For example, through virtual reality it may be possible to provide an opportunity to join a friend's space journey via a social media platform. The technology could also be used as a practical tool to train future spacecraft pilots, to educate students about space travel and its impacts on the environment or even be promoted as a sustainable way of experiencing space – from the comfort of your own sofa.

The era of the new space economy has brought new private actors to the formerly government monopolised arena and, aside from space tourism, there is vast economic potential in future space mining and private satellites supporting the technical functionality of the Earth. There is a new acceptance of the inevitable, needing to be managed appropriately to react with the continuously changing trends in the environment, technological achievements and global politics, in determining the future direction of space planning and legislation. It is essential to establish new space economy regulations to advise on how to operate in a sustainable manner and emphasise that environmental assessments should always be undertaken prior to any new development of space activities. Space technology projects like the clearing of space debris and using solar wind sails for sustainable space transportation should be legally enhanced with, for instance, tax relief. The absence of planning or valid legislation, which does not anticipate a sustainable future, could otherwise result in serious inefficiencies for future generations.

The pioneering years of the space tourism industry will show whether sustainability is seriously considered and acted upon to ensure a more sustainable future for this new sector of tourism and its impacts on society. According to Spector (2020), at the moment the simple fact is that space tourism sustainability is still suffused with many 'ifs', with many yet unknown future factors. However, it can be argued that if the tourism industry expands, and if this leads to reduced costs and facilitates space-based solar power, and if that actually works, and if the energy is shared equitably around the world – or – if space tourism leads to reduced launch costs, and if asteroid mining actually works, and if for some inexplicable reason the wealth available in space leads to peace instead of warfare (contrary to former actions in human history) – then space tourism has contributed to sustainability.

Also, as space tourism will initially start as an elite activity, there is a need to carefully evaluate whether, again in light of human history, the impacts of such elite activities will help to solve issues plaguing human society. Or whether we will ultimately find ourselves in need of ever-increasing resources, added to by the growth in space tourism transportation to future colonies, unable to revert back to a more sustainable and low-energy way of living on Earth (Spector, 2020).

To reflect some of the public views on sustainable space tourism, the future space tourism industry could learn from aviation that the only way to achieve sustainable growth is to supply services that the public want to buy. Firstly, passenger safety must be secured and safeguarded so that the image of space tourism does not immediately suffer from an unsafe future image. Secondly, the image of the space tourism industry is one of increased emissions and the creation of more space debris in this time of climate crisis, with a limited display of attention paid to ecological and ethical considerations, even those who could afford a trip may pass up the experience to avoid potentially being shamed, instead of admired for their bravery, and this would certainly not be economically sustainable for the industry in the long run. The future of the space tourism industry is dependent on how the public views its value as it begins fully operating – the options are that it could remain nothing more than a dinner party topic for a small, elite group, or alternatively, perhaps only the universe will be our limit.

There are many angles to consider in the future sustainability of space tourism, and so the wise words of Stephen Hawking leave us with some final thoughts as we reflect upon the realities of the future on Earth:

> We face a number of threats: nuclear war, global warming and genetically engineered viruses. Although the chance of disaster on planet Earth in a given year may be quite low, it adds up over time, becoming a near certainty in the next thousand or ten thousand years. By that time, we should have spread out into space and to other stars, so it would not mean the end of the human race. However, we will not establish self-sustaining colonies in space for at least the next hundred years, so we have to be very careful in this period. (Stephen Hawking, 2016)

References

Abitzsch, S. (1996) Prospects of space tourism. European Aerospace Congress: Visions and limits of long-term aerospace developments. 15 May. See http://www.spacefuture.com/archive/prospects_of_space_tourism.shtml (accessed 19 October 2019a).

Aguilar, F.J. (1967) *Scanning the Business Environment*. New York: The Macmillan Company.

Alston, P. (2019) Climate change and poverty. UN Human Rights Council. See https://digitallibrary.un.org/search?f1=author&as=1&sf=title&so=a&rm=&m1=e&p1=UN.%20Human%20Rights%20Council.%20Special%20Rapporteur%20on%20Extreme%20Poverty%20and%20Human%20Rights&ln=en (accessed 2 August 2019).

Anderson, E. (2005) *Space Tourist's Handbook*. Philadelphia, PA: Quirk Books.

Anderson, K.L. and Bows, A. (2011) Beyond 'dangerous' climate change: Emission scenarios for a new world. *Philosophical Transactions of the Royal Society of London* 369 (1934), 20–44.

Ansoff, I.H. (1975) Managing strategic surprise by response to weak signals. *California Management Review* XVIII (2), 21–33.

Apollo (2019) The Project Apollo archive. See http://www.apolloarchive.com/ (accessed 17 October 2019).

Armstrong, N. (1969) Quote. In J.R. Hansen (2005) *First Man: The Life of Neil Armstrong*. New York: Simon & Schuster.

Ashford, D. (2002) *Spaceflight Revolution*. London: Imperial College Press.

Ashford, D. (2009) An aviation approach to space transportation. *The Aeronautical Journal* 113 (1146), 499–515.

Astro Society (2019) K-12 science teachers. See https://astrosociety.org/education-outreach/k-12-science-teachers/project-astro.html (accessed 16 November 2019).

Baptista, J.A. (2014) The ideology of sustainability and the globalisation of a future. *Time & Society* 23 (3), 358–379.

Barr, S. (2008) *Environment and Society: Sustainability, Policy and the Citizen*. Aldershot: Ashgate.

BBC (2007) Hawkings takes zero gravity flight. *BBC News*. 27 April.

Beard, S. and Starzyk, J. (2002) *Space Tourism Market Study: Orbital Space Travel & Destinations with Suborbital Space Travel*. Bethesda, MD: Futron Corporation. See https://www.spaceportassociates.com/pdf/tourism.pdf (accessed 1 August 2019).

Beck, J. (2016) Space travel with VR. See http://www.virtual-reality-in-tourism.com/space-travel-with-vr/ (accessed 17 November 2019).

Becken, S. (2015) *Tourism and Oil: Preparing for the Challenge*. Bristol: Channel View Publications.

Beedie, P. and Hudson, S. (2003) Emergence of mountain-based adventure tourism. *Annals of Tourism Research* 30, 625–643.

Bell, W. (1997) *Foundations of Future Studies. Vol. 1-11.* New Brunswick, NJ: Transaction Books.

Benn, S., Dunphy, D. and Griffiths, A. (2006) Enabling change for corporate sustainability: An integrated perspective. *Australasian Journal of Environmental Management* 13 (3), 156–165.

Bezos, J. (2019) Quote in Hamilton, I.A. (2019) Jeff Bezos says space travel is essential because we are "in the process of destroying this planet". *Business Insider*, 18 July, https://www.businessinsider.com/jeff-bezos-space-travel-essential-because-destroying-planet-2019-7?r=US&IR=T (accessed 1 September 2020)

Bigelow Aerospace (2019) The Bigelow expandable activity module. See http://www.bigelowaerospace.com/pages/beam/ (accessed 21 October 2019).

Biosphere2 (2019) Biosphere2. Where the science lives. See http://biosphere2.org/ (accessed 29 October 2019).

Birnie, P.W. and Boyle, A.E. (1992) *International Law and the Environment.* Oxford: Clarendon Press.

Blue Origin (2019) Our mission. See https://www.blueorigin.com/our-mission (accessed 20 October 2019).

Blue Origin (2020) NASA selects Blue Origin National Team to return humans to the Moon. https://www.blueorigin.com/news/nasa-selects-blue-origin-national-team-to-return-humans-to-the-moon (accessed 3 August 2020).

Bluth, B.J. (1979) Sociology and space development, California State University. See https://er.jsc.nasa.gov/seh/sociology.html (accessed 29 October 2019).

Bradbury, D. (2010) Space tourism: A final frontier worth exploring? *The Guardian.* GSB Editorial Network. 23 October.

British Airways (2019) Corporate responsibility. See https://www.britishairways.com/en-gb/information/about-ba/csr/corporate-responsibility (accessed 2 November 2019).

Broderick, J. (2009) Voluntary carbon offsets: A contribution to sustainable tourism? In S. Gössling, C.M. Hall and D. Weaver (eds) *Sustainable Tourism Futures: Perspectives on Systems, Restructuring and Innovations* (pp. 169–199). New York: Routledge.

Bruntland Commission (1987) *Our Common Future.* Oxford: Oxford University Press.

Bunger, F. (2018) Orion span takes steps towards a giant leap in space tourism. See https://www.futuretravelexperience.com/2018/07/orion-span-takes-steps-towards-a-giant-leap-in-space-tourism/ (accessed 15 September 2018).

Butler, R.W. (1980) The concept of a tourist area cycle of evolution: Implications for management of resources. *Canadian Geographer/Le Geographe Canadien* 24, 5–12.

Byers, M. and Byers, C. (2017) Toxic splash: Russian rocket stages dropped in Arctic waters raise health, environmental and legal concerns. *Polar Record* 53 (6), 580–591.

Caplan, N., Winnard, A. and Lindsay, K. (2017) Here is what space tourism could do for science and health research. *The Conversation*, 22 June.

Carson, R. (1962) *Silent Spring.* Boston, MA: Houghton Mifflin.

Carter, C., Garrod, B. and Low, T. (2015) *The Encyclopaedia of Sustainable Tourism.* Boston, MA: CAB International.

Cater, C. (2019) History of space tourism. In E. Cohen and S. Spector (eds) *Space Tourism. The Elusive Dream.* Tourism Social Science Series Vol. 25 (pp. 51–66). Bingley: Emerald Publishing Limited.

Ceuterick, M. and Johnson, M.R. (2019) Space tourism in contemporary cinema and video games. In E. Cohen and S. Spector (eds) *Space Tourism. The Elusive Dream.* Tourism Social Science Series Vol. 25 (pp. 93–115). Bingley: Emerald Publishing Limited.

Chaikin, A. (2015) Is SpaceX changing the rocket equation? *Air & Space Magazine.*

Chhanivara, S. (2019) *The Future of Virtual Reality: Will We Soon Be Travelling Inside Our Heads.* Virgin: The Future of Travels Series. See https://www.virgin.com/travel/the-future-of-virtual-reality-will-we-soon-be-travelling-inside-our-heads (accessed 17 November 2019).

Choo, C.W. (2006) The art of scanning the environment, ASIS Bulletin Article Pre-print, *ASIS Bulletin* 25 (3), 13–19. See http://choo.fis.utoronto.ca/FIS/respub/ASISbulletin/ (accessed 27 November 2019).

Choo, C.W. (2007) Information life cycle of emerging issues. See http://choo.fis.utoronto.ca/ncb/es/EsinfoLC.html (accessed 28 November 2019).

Cockell, C.S. (2007) *Space on Earth. Saving Our World by Seeking Others*. London: Macmillan.

Coffman, B.S. (1997) Weak signal research, Part III: Sampling, uncertainty and phase shifts in weak signal evolution. *Journal of Transition Management*. MG Taylor Corporation.

Cofield, C. (2017) President Trump signs NASA authorisation bill. 21 March. See https://www.space.com/36154-president-trump-signs-nasa-authorization-bill.html (accessed 9 November 2019).

Cohen, E. (2017) The paradoxes of space tourism. *Tourism Recreation Research* 42 (1), 22–31.

Cohen, E. and Spector, S. (2019) *Space Tourism: The Elusive Dream*. Bingley: Emerald Publishing Limited.

Cohen, S. and Taylor, L. (1992) *Escape Attempts: The Theory and Practice of Resistance to Everyday Life*. London: Routledge.

Cole, S. (2015) Space tourism: Prospects, positioning and planning. *Journal of Tourism Futures* 1 (2), 131–140.

Collins, P. (1994) Potential demand for passenger travel to orbit. Construction engineering and operations in Space IV. *ASCHE* 1, 578–586.

Collins, P. (2003) Space Tourism Market Demand and the Transportation Infrastructure. AIA/ICAS Symposium, 'The Next 100 Years'. 17 July. Dayton, Ohio.

Collins, P. (2014) 'Renaissance Italia', held at Politecnico di Milano, May 8–9. In S. Cole (ed.) Space tourism: Prospects, positioning and planning. *Journal of Tourism Futures* 1 (2), 131–140.

Collins, P. and Autino, A. (2010) What the growth of a space tourism industry could contribute to employment, economic, growth, environmental protection, education, culture and world peace. *Acta Astronautica* 66, 1553–1562.

Cooper, C., Fletcher, J., Fyall, A., Gilbert, D. and Wanhill, S. (eds) (2008) *Tourism: Principles and Practice*. Edinburgh: Prentice Hall.

Cornog, R. (1956) Economics of rocket-propelled aeroplanes. *Aeronautical Engineering Review*. September.

Crutzen, P.J. and Stoermer, E.F. (2000) The 'Anthropocene'. The International Geosphere-Biosphere Programme. *Newsletter* 41, 17–18.

Cuhls, K., Ganz, W. and Warnke, P. (2014) *Foresight Process on Behalf of German Ministry of Education and Research. New Future Fields*. Karlsruhe: Fraunhofer Institute for Systems and Innovation Research.

Damjanov, K. and Crouch, D. (2019) Virtual reality and space tourism. In E. Cohen and S. Spector (eds) *Space Tourism. The Elusive Dream*. Tourism Social Science Series Vol. 25 (pp. 117–137). Bingley: Emerald Publishing Limited.

Davenport, C. (2015) Space tourism is closer to taking off, but how should it be regulated? *Los Angeles Times*. Business/Technology, 24 November.

David, L. (2014) China's new spaceport to launch country's largest rocket yet. *Space Insider*. 2 April. See https://www.space.com/25323-china-new-spaceport-rocket-launches.html (accessed 7 November 2019).

Davidson, J. (2019) Australia's fires emitted half of the country's annual CO_2 emissions. *Ecowatch*. 17 December. See https://www.ecowatch.com/australia-fires-co2-emissions-2641621994.html (accessed 15 January 2020).

Davies, G. (2013) Apprising weak and strong sustainability: Searching for middle ground. *The Journal of Sustainable Development* 10 (1), 111–124.

Dickens, P. (2019) Social relations, space travel, and the body of the astronaut. In E. Cohen and S. Spector (eds) *Space Tourism. The Elusive Dream.* Tourism Social Science Series Vol. 25 (pp. 51–66). Bingley: Emerald Publishing Limited.

Dobson, A. (1998) *Justice and the Environment: Conceptions of Environmental Sustainability and Dimensions of Social Justice.* Oxford: Oxford University Press.

Dornberger, W. (1942) Quote. See https://www.azquotes.com/author/42477-Walter_Dornberger (accessed 19 October 2019).

Dornberger, W.R. (1956) The Rocket Propelled Commercial Airliner. University of Minnesota, Institute of Technology, Research Report No. 135, November.

Dorozhkina, A. (2017) CosmoCourse will develop space tourism. 3 September 2019. See http://www.russia-ic.com/news/show/23821/#.Xd0Kw25uLIU (accessed 20 November 2019).

Dovers, S. and Handmer, J. (1992) Uncertainty, sustainability and change. *Global Environmental Change* 2 (4), 262–276.

Drake, N. (2018) They saw earth from space. Here is how IT changed them. *National Geographic.* See https://www.nationalgeographic.com/magazine/2018/03/astronauts-space-earth-perspective/ (accessed 4 November 2019).

Dredge, D. (2015) Tourism and covernance. In G. Moscardo and P. Benckendorff (eds) *Education for Sustainability in Tourism: A Handbook for Processes, Resources and Strategies* (pp. 75–90). New York: Springer Heidelberg.

Dredge, D. and Jenkins, J. (2007) *Tourism Policy and Planning.* Brisbane: Wiley.

Dredge, D. and Jenkins, J. (2011) *Stories of Practise: Tourism Planning and Policy.* Farnham: Ashgate.

Dryzek, J. (1997) *The Politics of the Earth.* Oxford: Oxford University Press.

Duval, D.T. and Hall, C.M. (2015) Sustainable space tourism. New destinations, new challenges. In C.M. Hall, S. Gössling and D. Scott (eds) *The Routledge Handbook of Tourism and Sustainability* (pp. 450–459). New York: Routledge.

Edelbroek, E. (2019) First baby could be born in space within 12 years. Space and Science Congress, Germany. See https://metro.co.uk/2019/10/16/first-baby-born-space-within-12years-expert-claims-10927087/ (accessed 22 November 2019).

EIA (2019) Transportation sector energy consumption. See https://www.eia.gov/outlooks/ieo/pdf/transportation.pdf (accessed 14 October 2019).

Environmentalism (2019) Definition. See https://www.merriam-webster.com/dictionary/environmentalism (accessed 15 October 2019).

ESA (2019a) Robotic exploration on Mars. See https://exploration.esa.int/web/mars/-/45801-european-heritage (accessed 29 October 2019).

ESA (2019b) Hibernating astronauts would need smaller spacecraft. See http://www.esa.int/Enabling_Support/Space_Engineering_Technology/Hibernating_astronauts_would_need_smaller_spacecraft (accessed 22 November 2019).

ESA (2020) Muscular-skeletal system: Bone and muscle loss. See https://www.esa.int/Enabling_Support/Preparing_for_the_Future/Space_for_Earth/Space_for_health/Musculo-skeletal_system_Bone_and_Muscle_loss (accessed 10 February 2020).

Etherington, D. (2020) Take a first look inside Virgin Galactic's spacecraft for private astronauts. https://techcrunch.com/2020/07/28/take-a-first-look-inside-virgin-galactics-spacecraft-for-private-astronauts/?guccounter=1&guce_referrer=aHR0cHM6Ly93d3cuZ29vZ2xlLmNvbS8&guce_referrer_sig=AQAAADCiKl-gPOCqDH8k-famCB5a0-l05ws-fOHw5l5uGM86QID-hxG783Dq-F8yN4i81o0tDVbRB0zXPvIu-JpN7f3uQOH8UL297XUsV5pEETbR6DeVNebQbs_dsBr1M_NNkTXtggWHSz-D24L18Grbede5qeq1LDLzQhDQRFmCajbVgJv (accessed 3 August 2020).

European Commission (2018) Progress made in cutting emissions. See https://ec.europa.eu/clima/policies/strategies/progress_en (accessed 3 November 2019).

Fawkes, S. (2007) Space tourism and sustainable development. *Journal of British Interplanetary Society* 60, 401–408.

FCC Aviation (2019) Swedish aviation tax. See https://centreforaviation.com/analysis/reports/aviation-emissions-swedens-flight-shame-possible-jet-fuel-tax-479859 (accessed 15 October 2019).

Federation Aeronautique Internationale – FAI (2018) Space Records. See https://www.fai.org/sport/space (accessed 10 December 2018).

Fingas, J. (2019) Russia will fly two space tourists to the ISS late 2021. See https://www.engadget.com/2019/02/20/russia-space-tourists-iss-2021/ (accessed 20 February 2019).

Finnair (2008) *Departure 2093. Viisi visiota lentomatkailusta*. Finnair: Paino Libris Oy.

Finnair (2020) Push for change. See https://www.finnairshop.com/en/sustainability-products (accessed 10 February 2020).

Fletcher, J. (2008) Sustainable tourism. In C. Cooper, J. Fletcher, A. Fyall, D. Gilbert and S. Wanhill (eds) *Tourism: Principles and Practice* (pp. 214–239). Edinburgh: Prentice Hall.

Fletcher, R., Murray, I., Blanco, A. and Salom Blazquetz, B. (2019) Tourism and degrowth: An emerging agenda for research and praxis. *Journal of Sustainable Tourism* 27 (12), 1745–1763.

FMI (2019) Steam balloon to facilitate satellite launches. Finnish Meteorological Institute. See https://en.ilmatieteenlaitos.fi/news/1105904583 (accessed 1 September 2019).

Fuad-Luke, A. (2010) *Eco-Design: The Sourcebook*. San Francisco, CA: Chronicle Books.

Gallego, J. (2018) The next stage of evaluation: How will the human species evolve. *Futurism*. 31 August. See https://futurism.com/the-next-stage-of-evolution-how-will-the-human-species-evolve (accessed 15 November 2019).

Gajanan, M. (2015) Virgin Galactic crash: Co-Pilot unlocked braking system too early. *The Guardian* https://www.theguardian.com/science/2015/jul/28/virgin-galactic-spaceshiptwo-crash-cause (accessed 3 August 2020).

García-Rosell, J.C. (2019) Vastuullinen matkailu. In J. Edelheim and H. Ilola (eds) *Matkailututkimuksen avainkäsitteet* (pp. 229-234). Rovaniemi: Lapland University Press.

Garrett-Bakeman, F.E. (2019) The NASA twin study: A multidimensional analysis of a year-long human space flight. *Science* 364, 6436.

GESY (2019) Global energy statistical yearbook 2019. See https://yearbook.enerdata.net/total-energy/world-consumption-statistics.html (accessed 15 October 2019).

Getz, D. (1986) Models in tourism planning: Towards integration of theory and practice. *Tourism Management* 7, 21–23.

Global Information (2018) Global space tourism market size, status and forecast 2025. *Market Research Report*. See https://giiresearch.com (accessed 20 October 2019).

Gössling, S., Peeters, P., Ceron, J.-P., Dubois, G., Patterson, T. and Richardson, R.B. (2005) The eco-efficiency of tourism. *Ecological Economics* 54, 417–434.

Gössling, S., Hall, C.M. and Weaver, D. (eds) (2009) *Sustainable Tourism Futures: Perspectives on Systems, Restructuring and Innovations*. New York: Routledge.

Gott, J.R. (2002) *Time Travel in Einstein's Universe: The Physical Possibilities of Travel Through Time*. London: Phoenix.

GovUK (2018) UK space industry sets out vision for growth. 11 May. See https://www.gov.uk/government/news/uk-space-industry-sets-out-vision-for-growth (accessed 10 October 2019).

Griffin, A. (2018) Travelling to Mars and deep into space could kill astronauts by destroying the guts. *Independent*. See https://www.independent.co.uk/news/science/nasa-mars-deep-space-journey-guts-gi-digestive-animal-study-gastrointestinal-health-a8563926.html (accessed 20 October 2019).

Grush, L. (2018) SpaceX will send Japanese billionaire Yasuku Maezawa to the moon. *The Verge*. 17 September. See https://www.spacex.com/news/2017/02/27/spacex-send-privately-crewed-dragon-spacecraft-beyond-moon-next-year (accessed 22 October 2019).

Gueguinou, N., Huin-Schohn, C., Bascove, M., Buelb, J.-L., Tschirhart, E., Legrand-Frossi, C. and Frippiat, J.-P. (2009) Could spaceflight associated immune system weakening preclude the expansion of human presence beyond earth's orbit. *Journal of Leukocyte Biology* 86 (5), 1027–1038.

Gunn, C.A. (1979) Resources management for visitors. *Fish and Wildlife News* 20 (October–November).

Gyimah, S. (2018) Quote. UK space industry sets out vision for growth. See https://www.ukspace.org/wp-content/uploads/2019/05/Prosperity-from-Space-strategy_2May2018.pdf (accessed 8 October 2019).

Hale, E.E. (1869) The brick moon. *Atlantic Monthly* Vol. 24.

Hall, C.M. (2011) Policy learning and policy failure in sustainable tourism governance: From first and second order to third-order change. *Journal of Sustainable Tourism* 19 (4–5), 649–671.

Harvey, D. (1989) *The Urban Experience*. Baltimore, MD: The John Hopkins University Press.

Hastings, J.G. (2014) The rise of Asia in a changing Arctic: A view from Iceland. *Polar Geography* 37 (3), 215–233.

Hawking, S. (2010) Stephen Hawking's warning: Abandon earth – or face extinction. See https://www.bbc.com/news/science-environment-43408961 (accessed 9 October 2019).

Hawking, S. (2016) What motivates Stephen Hawking? His answer has the power to inspire US all. *Radio Times*. Interview. 19 January. See https://web.archive.org/web/20180312182957/http://www.radiotimes.com/news/2016-01-19/what-motivates-stephen-hawking-his-answer-has-the-power-to-inspire-us-all/ (accessed 1 November 2019).

Heikkilä, M. (2018) Avaruusromu odottaa siivoojaa – Tilanne vastaa sitä, että käytöstä poistetut rekat jätettäisiin moottoriteiden varsille. *YLE*. Tiedeykkönen. 4 May. See https://yle.fi/aihe/artikkeli/2018/05/04/avaruusromu-odottaa-siivoojaa-tilanne-vastaa-sita-etta-kaytosta-poistetut-rekat (accessed 23 March 2019).

Heikkinen, S. (2018) Ilmastonmuutos on ihmiskunnan kohtalon kysymys. *YLE*. See https://yle.fi/aihe/artikkeli/2018/12/12/ilmastonmuutos-on-ihmiskunnan-kohtalonkysymys-tutki-kuka-paastoista-oikeastaan (accessed 18 October 2018).

Heininen, L. (2017) Environmental Impacts and Risks of the Military in Peacetime, Arctic Circle Assembly, October. Presentation, Reykjavik, Iceland.

HeliosTour (2020) Northern Lights in Rovaniemi. See http://aurora-rovaniemi.com/?gclid=CjwKCAjwmKLzBRBeEiwACCVihvqXaiV5rBwlIUuzwxi1frOW8srg9lHJsT2wVSIu1arBWkKVI_yL8RoC9SMQAvD_BwE (accessed 15 February 2020).

Helne, T. and Silvasti, T. (eds) (2012) *Yhteyksien kirja. Etappeja ekososiaalisen hyvinvoinnin polulla*. Tampere: Kelan tutkimusosasto.

Hickman, J. (1999) The political economy of very large space projects. *Journal of Evolution and Technology* 4, 1–14.

HIE (2019) Space Hub Sutherland. Highlands and Islands Enterprise. See https://www.hie.co.uk/our-region/regional-projects/space-hub-sutherland/ (accessed 2 October 2019).

Hiltunen, E. (2007) *Where Do Future Oriented People Find Weak Signals*. Turku: FFRC Publications.

Hiltunen, E. (2013) *Foresight and Innovation: How Companies Are Coping with the Future*. Basingstoke: Palgrave Macmillan.

Hiltunen, E. (2019) *Tulossa huomenna. Miten megatrendit muokkaavat tulevaisuuttamme*. Jyväskylä: Docendo.

Hubbert, M.K. (1956) Nuclear energy and the fossil fuels. *Drilling and Production Practice. American Petroleum Institute*. Houston: Shell Development Company.

Hubbert, K. (1962) Energy resources. *National Academy of Sciences* 1000-D, 81–83.

Hutton, G. (2019) When will UK spaceports be ready for lift-off? House of Commons. See https://commonslibrary.parliament.uk/science/technology/when-will-uk-spaceports-be-ready-for-lift-off/ (accessed 10 October 2019).

Höckert, E. (2015) *Ethics of Hospitality: Participatory Tourism Encounters in the Northern Highlands of Nicaragua.* Acta Universitatis Lapponiensis 312. Rovaniemi: Lapin yliopisto.

Hollingham, R. (2013) What will it take to set up colonies in space. BBC Future. See https://www.bbc.com/future/article/20131201-how-to-set-up-home-in-space (accessed 15 June 2019).

IEA (2019) Aviation: Tracking clean energy progress. See https://www.iea.org/topics/tracking-clean-energy-progress (accessed 14 October 2019).

Insinna, V. (2018) Trump's new space force to reside under Department of the Air Force. *Defence News.* See http://defencenews.com/ (accessed 20 December 2018).

IPCC (2018) Summary for Policymakers of IPCC Special Report on Global Warming of 1.5 Celsius Approved by the Governments. See https://www.ipcc.ch/2018/10/08/summary-for-policymakers-of-ipcc-special-report-on-global-warming-of-1-5c-approved-by-governments/ (accessed 1 November 2019).

IPCC (2019) The Intergovernmental Panel on Climate Change. See https://www.ipcc.ch/ (accessed 10 October 2019).

ISRO (2019) About ISRO. Department of space, Indian space research Organisation. See https://www.isro.gov.in/ (accessed 2 November 2019).

Janhunen, P. (2004) Electric sail for spacecraft propulsion. *Journal of Propulsion and Power* 20 (4) 763–764.

Janhunen, P. (2019) Recent research: Steam balloon to facilitate satellite launches. Finnish Meteorological Institute. 28 August. See https://en.ilmatieteenlaitos.fi/news/1105904583 (accessed 10 December 2019).

Jensen, R. (1999) *Dream Society. How the Coming Shift from Information to Imagination Will Transform Your Business.* New York: McGraw-Hill.

Johnston, I. (2017) Thousands of people could live in space colonies orbiting the earth in 20 years, expert claims. *Independent.* 11 March. See https://www.independent.co.uk/news/science/space-colonies-orbiting-earth-20-years-expert-prediction-a7623726.html (accessed 8 October 2019).

Kahn, H. (1965) *Thinking about the Unthinkable.* New York: Horizon Press.

Kahn, H. and Wiener, A. (1967) *The Year 2000. A Framework for Speculation on the Next Thirty-Three Years.* New York: MacMillan.

Kanas, N. and Manzey, D. (2008) Basic issues of human adaptation to space flight. *Space Psychology and Psychiatry* 22, 15–48.

Kennedy, J.F. (1961) The decision to go to the Moon. NASA archives. See https://history.nasa.gov/moondec.html (accessed 2 October 2019).

Kennedy-Pipe, C. (2017) Whose Security? The Arctic, Its People and Its Resources, Arctic Circle Assembly, October. Presentation, University of Hull, Reykjavik, Iceland.

Kennedy Space (2019) Kennedy Space Centre visitor complex. See https://www.kennedyspacecenter.com/ (accessed 10 November 2019).

Keränen, M. (2019a) SpaceX laukaisi 60 uutta laajakaistasatelliittia avaruuteen. *Tekniikka & Talous.* Uutiset. 13 November. Alma Talent, Finland.

Keränen, M. (2019b) Tämä arvio on vanhoille avaruusyhtiöille karvasta luettavaa: SpaceX pyyhkii Boeingilla pöytää. *Tekniikka ja Talous.* 15 November. Alma Talent, Finland.

Khan, H., Seng, C. and Cheong, W. (1990) Tourism multiplier effects on Singapore. *Annals of Tourism Research* 17, 408–418.

Khan, M. (1995) Sustainable development: The key concepts, issues and implications. *Sustainable Development* 3 (2), 63–69.

Kivimäki, A. (2019) NASAn teknologiajohtaja: Kuun kautta mennään Marsiin. *Helsingin Sanomat.* Tiede. 19 September. Sanoma, Finland.

Koistinen, A. (2018) Ihmiskunnan ratkaisevat vuodet. *YLE*. See www.yle.fi/uutiset (accessed 8 October 2018).

Kopra, S. (2017) China, Climate Change and International Security; Changing Attributes of Great Power Responsibility, Arctic Circle Assembly, October. Presentation, Reykjavik, Iceland.

Kosow, H. and Gassner, R. (2008) *Methods of Future and Scenario Analysis. Overview, Assessment and Selection Criteria*. Bonn: Deutches Institut fur Entwicklungspolitik.

Krippendorf, K. (1994) Redesigning design. In P. Tahkokallio and S. Vihma (eds) *Design: Pleasure or Responsibility* (pp. 138–162). Helsinki: University of Art and Design.

Kuusi, O. (2017) The Delphi method. In O. Kuusi, S. Heinonen and H. Salminen (eds) *How Do We Explore Our Futures? Methods of Futures Research*. Acta Futura Fennica 10. The Finnish Society for Future Studies.

Kuusi, O., Cuhls, K. and Steinmuller, K. (2015) The futures map and its quality criteria. *European Journal of Futures Research* 3 (22), 1–14.

Lane, B. (2009) Thirty years of sustainable tourism. Drivers, progress, problems – and the future. In S. Gössling, C.M. Hall and D. Weaver (eds) *Sustainable Tourism Futures. Perspectives on Systems, Restructuring and Innovations* (pp. 19–31). New York: Routledge.

Launius, R.D. (2019) Human aspirations to expand into space: A historical review. In E. Cohen and S. Spector (eds) *Space Tourism. The Elusive Dream*. Tourism Social Science Series Vol. 25 (pp. 15–50). Bingley: Emerald Publishing Limited.

Lee, R. (2003) Costing and Financing a Commercial Asteroid Mining Venture. Fifty-Fourth International Astronautical Congress, Bremen, Germany.

Legislation UK (2018) Space Industry Act 2018. See http://www.legislation.gov.uk/ukpga/2018/5/contents/enacted (accessed 10 October 2019).

Lele, S. (1991) Sustainable development: A critical review. *Word Development* 19 (6), 607–621.

Lewis, S. (1997) *Mining the Sky: Untold Riches from the Asteroids, Comets and Planets*. Reading, MA: Addison-Wesley.

Ma, C. (2017) China aims to be world-leading space power by 2045. *ChinaDaily*, 17 November. See https://www.chinadaily.com.cn/china/2017-11/17/content_34653486.htm (accessed 3 March 2019).

Ma, M. and Hassink, R. (2013) An evolutionary perspective on tourism area development. *Annals of Tourism Research* 41, 89–109.

MacDonald, A. (2017) *The Long Space Age: The Economic Origins of Space Exploration from Colonial America to the Cold War*. New Haven, CN: The Yale University Press.

Malm, A. and Hornborg, A. (2014) The geology of mankind? A critique of the Anthropocene narrative. *The Anthropocene Review* 1 (1), 62–69.

Mansi (2019) NASA to open space tourism for private trips. *Business Technology*, 9 June. See https://www.blocktoro.com/p/nasa-to-open-space-tourism-for-private-trips/ (accessed 11 November 2019).

Markley, O. (2011) A new methodology for anticipating STEEP surprises. *Technological Forecasting and Social Change* 78, 1079–1097.

Marsh, M. (2006) Ethical and medical dilemmas of space tourism. *Science Direct* 37 (9), 1823–1827.

Mathieson, A. and Wall, G. (2008) *Tourism: Economic, Physical and Social Impacts*. Harlow: Longman.

Matos, L.H. and Afsarmanesh, H. (2004) *Collaborative Networked Organizations: Research Agenda for Emerging Business Models*. Boston, MA: Kluwer Academic.

Matthews, R. (2019) Space Sustainability ASRC19 Conference Reprint. See https://www.researchgate.net/publication/336130754_Space_Sustainability_ASRC19_Conference_Preprint (accessed 10 February 2020).

McFadden, M. (2018) SpaceX and the ethics of space travel. *Prindle Post*. 6 February. See https://www.prindlepost.org/2018/02/spacex-ethics-space-travel/ (accessed 15 October 2019).

McGrath, M. (2019) Climate change: Trees most effective solution for warming. BBC Science and Environment. 4 July. See https://www.bbc.com/news/science-environment-48870920 (accessed 2 November 2019).

McKercher, B. (2005) Are psychographics predictors of destination life cycles? *Journal of Travel and Tourism Marketing* 19 (1), 49–55.

McKnight, J.C. (2003) The space settlement summit. *Space Daily*. 20 March.

MDRS (2019) Mars Desert Research Station. See http://mdrs.marssociety.org/about/ (accessed 29 October 2019).

Meadows, D. and Randers, J. (2005) *Limits to Growth: The 30-Year Update*. London: Earthscan.

MiGFlug (2019) Be a fighter pilot for a day. See https://migflug.com/flights-prices/mig-29-edge-of-space/ (accessed 23 October 2019).

Milano, C., Novelli, M. and Cheer, J.M. (2019) Overtourism and degrowth: A social movements perspective. *Journal of Sustainable Tourism* 27 (12), 1857–1875.

Mittal, R. (2013) Impact on population explosion on environment. *Weschool Knowledge Builder – The National Journal* 1 (1).

Mojave (2019) Mojave Air & Space Port. See https://www.mojaveairport.com/ (accessed 1 November 2019).

Mojic, J. and Susic, V. (2014) Planning Models of Sustainable Development Destination. International Scientific Conference: The Financial and Real Economy: Towards Sustainable Growth, University of Nis, 17 October.

Moon, M. (2018) Blue Origin's suborbital flights might cost 200,000 dollars per ticket. 13 July. See https://www.engadget.com/2018/07/13/blue-origin-suborbital-flight-ticket-price (accessed 1 October 2019).

Moore, J.W. (2014) The Capitalocene Part I: On the nature and origins if our ecological crisis. See https://www.researchgate.net/publication/264707683_The_End_of_Cheap_Nature_or_How_I_Learned_to_Stop_Worrying_about_'the'_Environment_and_Love_the_Crisis_of_Capitalism (accessed 10 March 2019).

Moore, J.W. (2015) *Capitalism in the Web of Life: Ecology and the Accumulation of Capital*. New York: Verso Books.

Morgan, G. (2006) *Images of Organization*. London: SAGE.

Mouyal, L.W., Morgensen, B. and Jingling, S. (2016) Greenland and Chinese outbound investments. *Advances in Polar Science* 27 (3), 192–199.

Mowforth, M. and Munt, I. (1998) *Tourism and Sustainability: New Tourism in the Third World*. London: Routledge.

Mowforth, M. and Munt, I. (2015) *Tourism and Sustainability: Development, Globalisation and New Tourism in the Third World* (4th edn). Abingdon: Routledge.

Musk, E. (2017) Making humans a multi-planetary species. *New Space* 2 (2). See https://www.liebertpub.com/doi/full/10.1089/space.2017.29009.emu (accessed 2 November 2019).

Musk, E. (2019) Making life multiplanetary. See https://www.spacex.com/mars (accessed 21 October 2019).

Närhi, K. (2004) The eco-social approach in social work and the challenges to the expertise of social work. Academic dissertation, University of Jyväskylä.

NASA (2019a) Humans in space. See https://www.nasa.gov/topics/humans-in-space (accessed 15 July 2019).

NASA (2019b) Human research program. See https://www.nasa.gov/twins-study (accessed 20 October 2019).

NASA (2019c) International Space Station. See https://www.nasa.gov/mission_pages/station/main/index.html (accessed 22 October 2019).

NASA (2019d) Mars exploration program. See https://mars.nasa.gov/#mars_exploration_program/0 (accessed 22 October 2019).

NASA (2019e) Space tourism posters. See https://solarsystem.nasa.gov/resources/682/space-tourism-posters/ (accessed 23 October 2019).

NASA (2019f) NASA opens international space station to new commercial opportunities. See https://solarsystem.nasa.gov/resources/682/space-tourism-posters/ (accessed 23 October 2019).

NASA (2019g) New space policy directive calls for human expansion across solar system. See https://www.nasa.gov/press-release/new-space-policy-directive-calls-for-human-expansion-across-solar-system (accessed 20 October 2019).

NASA (2019h) Space colony. See https://www.hq.nasa.gov/office/hqlibrary/pathfinders/colony.htm (accessed 30 October 2019).

NASA (2019i) Mars facts. See https://mars.nasa.gov/all-about-mars/facts/ (accessed 1 November 2019).

NASA (2019j) Space resources. See https://isru.nasa.gov/SPACERESOURCES.html (accessed 10 November 2019).

NASA (2019k) Planetary protection. See https://sma.nasa.gov/sma-disciplines/planetary-protection (accessed 10 November 2019).

NASA (2019l) Baikonur Cosmodrome. See https://www.nasa.gov/mission_pages/station/structure/elements/baikonur.html (accessed 10 November 2019).

NASA (2019m) With Mars methane mystery unsolved, curiosity serves scientists a new one: Oxygen. See https://www.nasa.gov/feature/goddard/2019/with-mars-methane-mystery-unsolved-curiosity-serves-scientists-a-new-one-oxygen (accessed 23 November 2019).

NASA (2020a) NASA astronauts launch from America in historic test flight of SpaceX Dragon Crew. https://www.nasa.gov/press-release/nasa-astronauts-launch-from-america-in-historic-test-flight-of-spacex-crew-dragon/ (accessed 1 August 2020).

NASA (2020b) NASA selects first commercial destination module for International Space Station. See https://www.nasa.gov/press-release/nasa-selects-first-commercial-destination-module-for-international-space-station (accessed 1 February 2020).

NASA (2020c) Shielding against space rays. See https://www.nasa.gov/audience/foreducators/5-8/features/F_Shielding_Space_Rays.html (accessed 10 February 2020).

NASDAQ (2020) Virgin Galactic holdings could be a home run stock. See https://www.nasdaq.com/articles/virgin-galactic-holdings-could-be-a-home-run-stock-2020-02-09 (accessed 9 February 2020).

Neergard, L. and Birenstein, S. (2019) Year in space put US astronauts disease defences of alert. *AP News*. See https://en.wikipedia.org/wiki/Effect_of_spaceflight_on_the_human_body (accessed 20 October 2019).

Newsela (2016) Astronauts eyes are at risk after too much time in space. *Science Washington Post*. See https://newsela.com/read/space-eyeballs/id/24554/ (accessed 20 October 2019).

Neymayer, E. (2003) *Weak Versus Strong Sustainability: Exploring the Limits of Two Opposing Paradigms*. London: Elgar.

Nurminen, P. (2019) Kiina sai siemenet itämään kuun pinnalla. *Iltalehti*. Ulkomaat. 15 January. Finland.

O'Kane, S. (2020) SpaceX will send three tourists to the International Space Station next year. See https://www.theverge.com/2020/3/5/21166657/spacex-tourists-iss-international-space-station-orbit-falcon-9-dragon (accessed 6 March 2020).

O'Neill, G.K. (1976) *The High Frontier: Human Colonies in Space*. New York: William Morrow & Company.

Ormond, J. and Dickens, P. (2019) Space tourism, capital and identity. In E. Cohen and S. Spector (eds) *Space Tourism. The Elusive Dream*. Tourism Social Science Series Vol. 25 (pp. 51–66). Bingley: Emerald Publishing Limited.

Osida (2019) Gateway to space. Oklahoma space industry development authority. See https://airspaceportok.com/ (accessed 2 November 2019).

Outer Space Treaty (1967) Treaty of Principles (UN). See http://www.unoosa.org/oosa/en/ourwork/spacelaw/treaties/introouterspacetreaty.html (accessed 17 September 2019).

Palmroth, M. (2018) Interview. In M. Heikkilä (ed.) Avaruusromu odottaa siivoojaa – Tilanne vastaa sitä, että käytöstä poistetut rekat jätettäisiin moottoriteiden varsille. *YLE.* Tiedeykkönen. 4 May. See https://yle.fi/aihe/artikkeli/2018/05/04/avaruus-romu-odottaa-siivoojaa-tilanne-vastaa-sita-etta-kaytosta-poistetut-rekat (accessed 21 March 2019).

Paukku, T. (2019a) Pikavuoro kuuhun peruttu – toistaiseksi. *Helsingin Sanomat.* Tiede. 13 March. Sanoma, Finland.

Paukku, T. (2019b) Yhdysvallat palaa kuuhun ennen vuotta 2029. *Helsingin Sanomat.* Tiede. 19 March. Sanoma, Finland.

Paukku, T. (2019c) Avaruuspurjeet ylös ja kohti lähitähti Proximaa. *Helsingin Sanomat.* Tiede. 20 August. Sanoma, Finland.

Pauwels, L. and Berger, J. (1964) *The Morning of the Magicians.* New York: Harper Torchbooks.

Peeters, P. (ed.) (2007) *Tourism and Climate Change Mitigation. Methods, Greenhouse Gas Reductions and Policies.* Breda: NHTV Academic Studies.

Peeters, P. (2018) Why space tourism will not be part of sustainable tourism. *Tourism Recreational Research* 43 (4), 540–543.

Peljo, J. (2018) Interview. In S. Heikkinen (2018) Ilmastonmuutos on ihmiskunnan kohtalon kysymys. *YLE.* See https://yle.fi/uutiset/3-10508539 (accessed 18 October 2018).

Peltoniemi, T. (2018) Ihminen haluaa levittäytyä avaruuteen. *YLE.* 2 November. See https://yle.fi/aihe/artikkeli/2018/11/02/ihminen-haluaa-levittaytya-avaruuteen?ref=ohj-articles (accessed 25 September 2019).

Perlman, D. (2009) NASA's moon blast called a smashing success. *The San Francisco Chronicle.*

Pincus, R. (2017) Emergency Response, International Relations and Security for Maritime and Coastal Tourism. Arctic Circle Assembly, October. Panellist from US Coast Guard Academy. Reykjavik, Iceland.

Plog, S. (1974) Why destination areas rise and fall in popularity? *Cornell Hotel and Administration Quarterly* 14, 55–58.

Praaks, J. (2018) Interview. In Avaruus avautui myös pienille maille – Suomella on satelliitteja, Ruotsilla pian rakettejakin Avaruustekniikka. *YLE.* See https://yle.fi/uutiset/3-10540615?fbclid=IwAR00NPSbPXH-XzuKWbazmWUvbrOR_mvzgHD8-9L3Oj7lbZcZNxm04WwgEIY (accessed 12 October 2019).

Quine, T. (2020) Orbital space tourism set for rebirth in 2021. https://www.thespacereview.com/article/4003/1 (accessed 10 August 2020).

Rametsteiner, E., Puelzl, H., Alkan-Olsson, J. and Fredriksen, P. (2011) Sustainable indicators: Science or political negotiation. *Ecological Indicators* 11 (1), 61–70.

Redd, N. (2018) Yuri Gagarin: First man in space. See https://www.space.com/16159-first-man-in-space.html (accessed 17 October 2019).

Rikard, L.M. and Borch, K. (2011) From future scenarios to roadmapping. Published in the 4th International Seville Conference on Future-Oriented Technology Analysis. DTU Library. See https://orbit.dtu.dk/files/6417528/Ricard2%20Brief%20from%20Future%20Scenarios%20to%20Roadmapping.pdf (accessed 29 October 2019).

Rincon, P. (2017) Queen's speech: Plan aims to secure space sector. See https://www.bbc.com/news/science-environment-40354695 (accessed 20 August 2017).

Ritchie, H. and Roser, M. (2018) Energy production and changing energy sources. *Our World in Data.* See http://ourworldindata.org/energy-production-and-changing-energy-sources (accessed 11 October 2019).

Rogers, J. (2017) Arctic Drones and Emerging Technologies. Arctic Circle Assembly, October. Presentation, University of York, Reykjavik, Iceland.

Roser, M. and Ortiz-Ospina, E. (2017) Global extreme poverty. Our world in data. See http://ourworldindata.org/extreme-poverty (accessed 10 October 2019).

Ross, M. and Sheaffer, P.M. (2014) Radiative forcing caused by rocket engine emissions. *Earth's Future* 2 (4), 177–196.

Ross, M., Mills, M. and Toohey, D. (2010) Potential climate impact of black carbon emitted by rockets. *Geophysical Research Letters* 37 (24), 1–6.

Ross, S. (2002) Near-earth asteroid mining. *Space Industry Report*. Blacksburg, VA: Virginia Polytechnic Institute and State University.

RT (2018) First ever luxury space hotel to be launched into orbit in 2021. 8 April. See https://www.rt.com/business/423518-first-luxury-space-hotel/ (accessed 13 November 2019).

RT (2019) Fly me to the moon: Russia may soon send tourists to space on private shuttle. 2 February. See https://www.rt.com/business/450384-tourists-russia-space-ship/ (accessed 14 November 2019).

Rudd, M., Aaker, J. and Vohls, K. (2012) Awe expands people's perception of time, alters decision making and enhances well being. *Psychological Science* 23 (10), 1130–1136.

Ryan, C. (2018) Future trends in tourism research: Looking back to look forward. The future of tourism management perspectives. *Tourism Management Perspectives* 25, 196–199.

Saarinen, J. (2014) Transforming destinations: A discursive approach to tourist destinations and development. In A. Viken and B. Granås (eds) *Tourism Destination Development, Turns and Tactics* (pp. 47–62). Farnham: Ashgate.

Saigol, L. (2019) As Virgin Galactic shares debut, Here are the billionaires leading the new space race. *Marketwatch*. See https://www.marketwatch.com/story/as-virgin-galactic-shares-debut-here-are-the-billionaires-leading-the-new-space-race-2019-10-28 (accessed 8 November 2019).

Salonen, A. (2010) *Kestävä kehitys globaalin ajan yhteiskunnan haasteena*. Helsinki: The University of Helsinki Research Centre.

Schmidt, S. and Zubrin, R. (eds) (1996) *Islands in the Sky: Bold New Ideas for Colonizing Space*. (pp. 71–84). New York: Wiley.

Seuri, V. (2018) Uusi jälleenrakennus vai sotatalous? *YLE*. Uutiset. See www.yle.fi/uutiset (accessed 7 October 2018).

Sharma, G.D. (2011) *Space Security: Indian Perspective*. New Delhi: Vij Books.

Sharpley, R. (2015) Sustainability: A barrier to tourism development. In R. Sharpley and D.J. Telfer (eds) *Tourism and Development: Concepts and Issues* (2nd edn; pp. 428–452). Bristol: Channel View Publications.

Sheetz, M. (2019a) Buy Virgin Galactic stock because space tourism will be safer than you think. *CNBC*. 5 November. See https://www.cnbc.com/2019/11/05/analyst-buy-virgin-galactic-stock-because-risk-is-misunderstood.html (accessed 9 November 2019).

Sheetz, M. (2019b) Superfast travel using outer space could be 20 billion dollar market, disrupting airlines. *CNBC*. 18 March. See https://www.cnbc.com/2019/03/18/ubs-space-travel-and-space-tourism-a-23-billion-business-in-a-decade.html (accessed 10 December 2019).

Sheetz, M. (2020) Virgin Galactic partners with Rolls-Royce as it looks to build an aircraft for supersonic air travel. *CNBC*. https://www.cnbc.com/2020/08/03/virgin-galactics-early-supersonic-aircraft-design-partnering-with-rolls-royce.html (accessed 3 August 2020).

Shepard, A. (2020) Quote. See https://www.quotetab.com/quotes/by-alan-shepard (accessed 12 February 2020).

Shevchuk, A. (2017) Evaluation and Elimination of Accumulated Ecological Damage in the Russian Arctic Zone. Arctic Circle Assembly, October. Presentation, Reykjavik, Iceland.

Smithsonian (2019) What does it mean to be human. See http://humanorigins.si.edu/evidence/human-fossils/species/homo-sapiens (accessed 31 October 2019).

Smyre, R. and Richardson, N. (2015) *Preparing for a World That Does Not Exist – Yet: Framing a Second Enlightenment to Create Communities of the Future*. New Alresford: John Hunt.

Space Adventures (2019) Zero gravity flight. See https://spaceadventures.com/experiences/zero-gravity-flight/ (accessed 23 October 2019).

Space Adventures (2020) Space adventures announces agreement with SpaceX to launch private citizens on the crew Dragon spacecraft. See https://spaceadventures.com/space-adventures-announces-agreement-with-spacex-to-launch-private-citizens-on-the-crew-dragon-spacecraft/ (accessed 17 February 2020).

Spaceflight101 (2019) Space news and beyond. See https://spaceflight101.com/2017-space-launch-statistics/ (accessed 10 November 2019).

Spaceport America (2019a) Virgin Galactic opens the doors to the 'Gateway to Space'. See https://www.spaceportamerica.com/virgin-galactic-opens-the-doors-to-the-gateway-to-space/ (accessed 2 November 2019).

Spaceport America (2019b) Visit us. See https://www.spaceportamerica.com/visit/ (accessed 10 November 2019).

Space Perspective (2020) A new perspective. See https://thespaceperspective.com/fly/ (accessed 14 July 2020).

Spaceport Sweden (2017) Your next adventure. See http://www.spaceportsweden.com/ (accessed 5 July 2017).

SpaceVR (2019) Space float now. See https://spacevr.co/ (accessed 17 November 2019).

SpaceX (2017) SpaceX to send privately crewed Dragon spacecraft beyond the moon next year. See https://www.spacex.com/news/2017/02/27/spacex-send-privately-crewed-dragon-spacecraft-beyond-moon-next-year (accessed 22 October 2019).

SpaceX (2019a) Making life multiplanetary. See https://www.spacex.com/mars (accessed 22 October 2019).

SpaceX (2019b) Capabilities and services. See https://www.spacex.com/about/capabilities (accessed 30 October 2019).

SpaceX (2020) About. See https://www.spacex.com/about (accessed 2 January 2020).

Spector, S. (2020) Personal communication. 12 March. Finland and New Zealand.

Spector, S. and Higham, J. (2019a) Space tourism in the Anthropocene. *Annals of Tourism Research* 79.

Spector, S. and Higham, J. (2019b) Space tourism, the Anthropocene and sustainability. In E. Cohen and S. Spector (eds) *Space Tourism. The Elusive Dream*, Tourism Social Science Series Vol. 25 (pp. 245–262). Bingley: Emerald Publishing Limited.

Spector, S., Higham, J.E.S. and Doering, A. (2017) Beyond the biosphere: Tourism, outer space and sustainability. *Tourism Recreational Research* 42 (3), 237–282.

Stallmer, E. (2015) Testimony of Eric W. Stallmer, President, Commercial Spaceflight Federation. Subcommitee on Science, Space and Competitiveness. 24 February. See http://www.commercialspaceflight.org/wp-content/uploads/downloads/2015/02/Eric-Stallmer-testimony.pdf (accessed 2 October 2019).

Stone, J. (2017) Thousands of people could live in space colonies orbiting the earth in 20 years, expert claims. In I. Johnstron. *Independent*. 11 March. See https://www.independent.co.uk/news/science/space-colonies-orbiting-earth-20-years-expert-prediction-a7623726.html (accessed 30 October 2019).

Strömberg, J. (2013) What is the Anthropocene and are we in it? See https://www.smithsonianmag.com/science-nature/what-is-the-anthropocene-and-are-we-in-it-164801414/ (accessed 25 September 2019).

STT (2007) Kiirunasta turistilennolle avaruuteen. See https://www.ts.fi/uutiset/maailma/1074176438/Kiirunasta+turistilennolle+avaruuteen (accessed 23 October 2019).

Stuart, G. (2018) Quote. In UK space agency sets out vision for growth. See https://www.ukspace.org/wp-content/uploads/2019/05/Prosperity-from-Space-strategy_2May2018.pdf (accessed 10 October 2019).

Sulasma, O-P. (2019) Venäjä laukaisi avaruusasemalle robonautin. *YLE. Uutiset.* See https://yle.fi/uutiset/3-10933364 (accessed 22 August 2019).

Sullivan, K. (1984) Spacewalk quote. In N. Drake (ed.) They saw earth from space. Here is how IT changed them. *National Geographic*. March 2018. See https://

www.nationalgeographic.com/magazine/2018/03/astronauts-space-earth-perspective/ (accessed 4 November 2019).

Sustainability 101 (2020) NASA Sustainability 101. See https://www.nasa.gov/emd/sustainability-101 (accessed 15 July 2020).

TEM (2020) Space offers new opportunities. The Ministry of Economic Affairs and Employment. See https://tem.fi/en/space (accessed 14 February 2020).

Tett, G. (2018) America unleashes billionaires to boost the space race. *Financial Times*, 1 June.

Toivonen, A. (2017) Sustainable planning for space tourism. *Matkailututkimus* 13 (1–2), 21–34.

Toivonen, A. (2019) Sustainable Space Tourism: Webpropol Survey. Survey conducted in Spring 2019. Helsinki, Finland.

Trump, D. (2017) New Space Policy Directive. See https://www.nasa.gov/press-release/new-space-policy-directive-calls-for-human-expansion-across-solar-system (accessed 20 October 2019).

Tuohinen, P. (2019) SpaceX:n kantoraketti tuhoutui räjähdyksessä Texasissa, koelaukaisu oli tarkoitus tehdä vielä tänä vuonna. *Helsingin Sanomat*. Tiede, 2 November. Finland: Sanoma.

UBS (2019) Space tourism: Ready for blast-off. *UBS*. See https://www.ubs.com/microsites/wma/insights/en/investing/2019/space-tourism.html (accessed 15 December 2019).

UK Space Agency (2020) New US-UK agreement boosts UK's spaceport plans. UK Space Agency. Press Release, 17 June, https://www.gov.uk/government/news/new-us-uk-agreement-boosts-uks-spaceport-plans (accessed 20 September 2020).

UN (1979) The Moon Treaty. See http://www.unoosa.org/oosa/en/ourwork/spacelaw/treaties/intromoon-agreement.html (accessed 11 October 2019).

UN (2015) The agenda for sustainable development. See https://sustainabledevelopment.un.org/post2015/transformingourworld (accessed 5 August 2020).

UN (2019a) World population prospects. See https://population.un.org/wpp/Download/Standard/Population (accessed 10 October 2019).

UN (2019b) The agenda for sustainable development. See https://sustainabledevelopment.un.org/content/documents/21252030%20Agenda%20for%20Sustainable%20Development%20web.pdf (accessed 15 October 2019).

UN (2019c) Social sustainability. See https://www.unglobalcompact.org/what-is-gc/our-work/social (accessed 9 November 2019).

UN (2019d) Outer Space Treaty. See http://www.unoosa.org/oosa/en/ourwork/spacelaw/treaties/introouterspacetreaty.html (accessed 10 October 2019).

UN Paris Agreement (2015) The Paris Agreement. See https://unfccc.int/process-and-meetings/the-paris-agreement/the-paris-agreement (accessed 10 September 2019).

UNWTO (2005) *Making Tourism More Sustainable: A Guide for Policy Makers*. Paris: United Nations Educational Programme.

Uskali, T. (2005) Paying attention to weak signals: The key concept for innovation journalism. *Innovation Journalism* 2 (11), 3–17.

21stcentech (2019) Will China, India and Japan join the space tourism race? 21st century tech: A look at our future. See https://www.21stcentech.com/china-india-japan-join-space-tourism-race/ (accessed 10 November 2019).

Vergano, D. (2014) Will Virgin Galactic's crash end space tourism? *National Geographic* https://www.nationalgeographic.com/news/2014/11/141031-virgin-galactic-space-tourism-impacts/ (accessed 3 August 2020).

Verbeek, D.H.P. and Mommaas, J.T. (2008) Transitions to sustainable tourism mobility: The social practises approach. *Journal of Sustainable Tourism* 16 (6), 557–565.

Vereshchetin, V., Vasilevskaya, E. and Kamenetskaya, E. (1987) *Outer Space. Politics and Law*. Moscow: Progress Publishers.

Viikari, L. (2007) *The Environmental Element in Space Law. Assessing the Present and Charting the Future*. Leiden: Brill Nijhoff.

Vince, G. (2016) *Adventures in the Anthropocene: A Journey to the Heart of the Planet We Made*. London: Penguin Random House.

Virgin Galactic (2019a) Virgin Galactic purpose. See https://www.virgingalactic.com/purpose/ (accessed 23 October 2019).

Virgin Galactic (2019b) Virgin Galactic vision. See https://www.virgingalactic.com/vision/ (accessed 22 October 2019).

Virgin Galactic (2020) Beth Moses: Reflections from an astronaut. See https://www.virgingalactic.com/articles/beth-moses-reflections-from-an-astronaut/ (accessed 14 February 2020).

Von der Dunk, F. and Tronchetti, F. (2015) *Handbook of Space Law. Research Handbooks in International Law*. Cheltenham: EE Elgar.

Von der Dunk, F. (2019) The regulation of space tourism. In E. Cohen and S. Spector (eds) *Space Tourism. The Elusive Dream*. Tourism Social Science Series (Vol. 25, pp. 177–199). Bingley: Emerald Publishing Limited.

Voshell, M. (2004) High acceleration and the human body. Research Gate. 28 November. See https://www.researchgate.net/publication/265032104_High_Acceleration_and_the_Human_Body (accessed 10 February 2020).

Waldek, S. (2020) 13 things space tourists should know before traveling to space, according to astronauts. *Travel & Leisure* https://www.travelandleisure.com/trip-ideas/space-astronomy/what-space-tourists-should-know-before-traveling-to-space-according-to-astronauts (accessed 2 August 2020).

Wall, M. (2011) First space tourist: How a US millionaire bought a ticket to orbit. 27 April 2018. See https://www.space.com (accessed 16 June 2018).

Wall, M. (2017) SpaceX's Mars colony plan. How Elon Musk plans to build a million person Martian city. See https://www.space.com/37200-read-elon-musk-spacex-mars-colony-plan.html (accessed 1 November 2019).

Wall, M. (2019a) Tickets to Mars will eventually cost less than 500,000 dollars, Elon Musk says. 13. February. See https://www.space.com/elon-musk-spacex-mars-mission-price.html (accessed 1 November 2019).

Wall, M. (2019b) European satellite dodge potential collision with SpaceX Starlink Craft. 3 September. See www.space.com/spacex-starlink-esa-satellite-collision-avoidance.html (accessed 14 November 2019).

Wallius, A. (2018) Avaruus avautui myös pienille maille – Suomella on satelliitteja, Ruotsilla pian rakettejakin. *YLE*. Avaruustekniikka. See https://yle.fi/uutiset/3-10540615?fbclid=IwAR00NPSbPXH-XzuKWbazmWUvbrOR_mvzgHD8-9L3Oj7l-bZcZNxm04WwgEIY (accessed 12 October 2019).

Weaver, D.B. (2005) *Sustainable Tourism: Theory and Practise*. Oxford: Elsevier.

Webber, D. (2010) Space Tourism: Essential Step in Human Settlement of Space. International Astronautical Federation 63rd International Astronautical Congress, Naples, Italy.

Webber, D. (2013) Space tourism: Its history, future and importance. *Acta Astronautica* 92 (2), 138–143.

Webber, D. (2019) Current space tourism developments. In E. Cohen and S. Spector (eds) *Space Tourism. The Elusive Dream* (pp. 163–175). Bingley: Emerald Publishing Limited.

Whitesides, G. (2019) Interview. In M. Sheetz. Buy Virgin Galactic Stock Because Space Tourism Will Be Safer Thank You Think, Analyst Says. *CNBC*. 5 November. See https://www.cnbc.com/2019/11/05/analyst-buy-virgin-galactic-stock-because-risk-is-misunderstood.html (accessed 10 November 2019).

Whoriskey, P. (2013) For Jeff Bezos, A New Frontier. *Washington Post*.

Wikipedia (2019a) Blue Origin. See https://en.wikipedia.org/wiki/Blue_Origin (accessed 23 October 2019).

Wikipedia (2019b) Space tourism. See https://en.wikipedia.org/wiki/Space_tourism (accessed 15 October 2019).

Wikipedia (2019c) Vostochny Cosmodrome. See https://en.wikipedia.org/wiki/Vostochny_Cosmodrome (accessed 5 November 2019).

Wikipedia (2019d) China Academy of launch vehicle technology. See https://en.wikipedia.org/wiki/China_Academy_of_Launch_Vehicle_Technology (accessed 7 November 2019).

Wikipedia (2019e) Tanegashima Space Center. See https://en.wikipedia.org/wiki/Tanegashima_Space_Center (accessed 7 November 2019).

Wilkinson, P.F. (1997) *Tourism Policy and Planning: Case Studies from the Commonwealth*. New York: Cognizant Communication.

Williams, V., Noland, R., Majumdar, A., Toumi, R. and Ochieng, W. (2007) Mitigation of climate impacts with innovative air transport management tools. In P. Peeters (ed.) *Tourism and Climate Change Mitigation. Methods, Greenhouse Gas Reductions and Policies* (pp. 91–104). Breda: NHTV Academic Studies.

Williamson, M. (2003) Space ethics and protection of the space environment. *Space Policy* 19 (1), 47–52.

Wilson, E. (2015) Practice what you teach: Teaching sustainable tourism through a critically reflexive approach. In G. Moscardo and P. Benckendorff (eds) *Education for Sustainability in Tourism. A Handbook for Processes, Resources and Strategies* (pp. 201–211). New York: Springer Heidelberg.

Wilson, D. (2018) Virtual space tourism: Travel tips for families and kids. See https://www.pandiapress.com/virtual-space-tourism-kids/ (accessed 17 November 2019).

Wittig, M.C., Beil, P., Sommerock, F. and Albers, M. (2017) *Rethinking Luxury: How to Market Exclusive Products and Services in an Ever-Changing Environment*. London: LID Publishing.

Wong, S. (2006) Sustainable water: Lessons from the developing world. *Engineering Sustainability* 159, 55–62.

Weise, K. (2020) Jeff Bezos commits 10 billion dollars to address climate change. *The New York Times*. Technology. See https://www.nytimes.com/2020/02/17/technology/jeff-bezos-climate-change-earth-fund.html (accessed 17 February 2020).

Yale Environment (2019) Planting 1.2 trillion trees could cancel out a decade of CO_2 emissions, scientists find. See https://e360.yale.edu/digest/planting-1-2-trillion-trees-could-cancel-out-a-decade-of-co2-emissions-scientists-find (accessed 15 October 2019).

Yamada, S. (2016) China aims for moon and Mars, rivalling US and Russia. See http://asia.nikkei.com/Business/Biotechnology/China-aims-for-moon-and-Mars-rivaling-US-and-Russia (accessed 23 November 2019).

Yeoman, I. (2008) *Tomorrow's Tourist: Scenarios and Trends*. Amsterdam: Elsevier.

Zero 2 Infinity (2020) Simplifying access to space. See http://www.zero2infinity.space/ (accessed 15 July 2020).

Zhukov, S. (2019) Russia's Cosmo Course launch private space tourism in five years. *The Power News*, 3 February. See https://etn.travel/russias-cosmocourse-could-launch-private-space-tourism-in-five-years-82455/#gsc.tab=0 (accessed 20 September 2019).

Index

Note: References in *italics* are to figures, those in **bold** to boxes.

For Product Safety Concerns and Information please contact our EU Authorised Representative:

Easy Access System Europe

Mustamäe tee 50

10621 Tallinn

Estonia

gpsr.requests@easproject.com

www.ingramcontent.com/pod-product-compliance
Lightning Source LLC
Chambersburg PA
CBHW050514280326
41932CB00014B/2317